3·1·95

D0857030

RF Radiation Safety Handbook

RF Radiation Safety Handbook

RONALD KITCHEN

Butterworth-Heinemann Ltd
Linacre House, Jordan Hill, Oxford OX2 8DP

℞ A member of the Reed Elsevier group

OXFORD   LONDON   BOSTON
MUNICH   NEW DELHI   SINGAPORE   SYDNEY
TOKYO   TORONTO   WELLINGTON

First published 1993

British Library Cataloguing in Publication Data
A catalogue record for this book is available from the British Library

Library of Congress Cataloguing in Publication Data
A catalogue record for this book is available from the Library of Congress

ISBN 0 7506 1712 8

Composition by Genesis Typesetting, Laser Quay, Rochester, Kent
Printed and bound in Great Britain by Redwood Books, Trowbridge

Contents

Preface

Are our brains being damaged by radio telephones? Can radio frequency signals cause our hair to fall out? These and similar assertions are the sort of comments which can be found in the media regularly. Anything that is unseen and mysterious creates some form of unease and we have to confess that none of these suggestions can be positively refuted or proven. Of course we can say that we do not believe them or that we have no evidence to support them but a lack of evidence is always open to the suggestion that we have not investigated the matter properly.

Nor is this problem new, as I recall reading that around the end of the nineteenth century, a man with a gun confronted a Marconi engineer on a transmitter site, claiming that he was being affected by the radiation.

This book is not so much about the advances in our knowledge of the effects of radio frequency (RF) radiation on human beings, though the hard work of researchers in the field should not be overlooked, but rather about the use of our current knowledge to safeguard people from any harmful effects of RF radiation. The basis used for this is the consideration of past and present RF radiation safety standards and their application to safety management.

In doing so we have to rely on the work of expert groups of people who have attempted to keep standards in line with current findings on RF safety. Not all the aspects of RF radiation which can cause harm relate directly to the human body. The potential hazards of RF radiation in respect of flammable vapours and electro-explosive devices are discussed and some mention is made of electro-magnetic compatibility (EMC) problems.

Since the book is addressed to people responsible for or concerned with safety, it cannot be assumed that all those involved are radio engineers. Often people from other disciplines such as mechanical engineering, chemistry and medicine may be involved. Consequently some attempt has been made to explain things which the radio engineer might consider everyday matters.

The general introduction to the book covers some of these aspects and is followed by pictorial examples of the sort of RF radiation sources likely to be met in RF radiation work, including RF process machines. Chapter 3 attempts to summarise the main aspects of those effects of RF radiation on people which are known or believed to have been established by research to date, together with references to sources of more detailed scientific treatment of the subject.

Chapter 4 discusses RF radiation standards from 1982 to the present time so that the development of ideas resulting from research can be illustrated. Detailed information on the limit values given in each standard are provided in the accompanying tables.

Chapter 5 deals with simple calculations for antenna fields ranging from microwave antennas to rod and whip antennas, together with other calculations involved in RF radiation survey work. This is followed by a chapter on RF radiation measuring instruments with practical examples which include some instruments which are currently being developed.

Because transmitters and other RF sources which use voltages above 5 kV may be subject to regulations governing X-ray radiation, Chapter 7 deals with X-ray production and measurement and the measuring instruments used. The hazards of this form of ionising radiation are also mentioned.

Chapter 8 deals with the preparations for RF and X-ray survey measurements and with the safety measures applicable and Chapter 9 is devoted to the practical methods involved in carrying out measurements. Chapter 10 then looks at the problem of designing equipment to provide for safe use in respect of RF and X-ray radiation.

The final chapter deals with the task of RF radiation safety management including training and RF radiation incident investigation. The book is meant to be essentially a practical one for those who have to be actively involved in RF safety work and I have tried to reflect in it some of the topics which have arisen in training mature students at the Marconi College.

At the time of writing the situation on standards is still very fluid with some standards and guides currently being reviewed or created. However I hope that the background in this book will help in understanding new standards, their basis and the reason for any changes in the limit values. A list of references to documents is included and some of the major reports mentioned such as those from the NRPB have their own extensive bibliographies.

I am indebted to so many people that it is difficult to know how to do justice to all of them. I am grateful to John Coulston, former Director of the GEC-Marconi Research Centre for permission to publish the material in Chapter 5 from the *Marconi Review*, for the use of the Research Library at Great Baddow and the help of the staff there, especially Arthur Jones and Pam Baxter.

I am similarly indebted to the British Standards Institution for permission to quote from several standards, to the USA Institute of Electrical and Electronic Engineers (IEEE) for permission to quote freely from the ANSI 1982 and the IEEE 1991 RF radiation standards, to the UK NRPB and NPL for help with information and pictures and to the World Health Organisation European Office for permission to quote from their material.

With regard to the manuscript I am grateful to former colleagues at the Marconi Research Centre, Steven Phillipo and Dr Douglas Shinn, for reading and commenting on some chapters, and to Dr K. C. Edsall (Wing Commander, Royal Air Force; Head of Radiation Medicine at DRPS) for commenting on the medical section. Needless to say, I remain responsible for any errors remaining.

Many equipment suppliers have assisted with photographs and technical information on their equipment and this includes the GEC-Marconi Radar and Communications Companies in Chelmsford. Other contributions include the Civil Aviation Authority and Mercury Communications.

During the course of writing the book I have had much personal support from colleagues on the Ministry of Defence Industrial Liaison Group RF Working Group which is a group involving most organisations in the UK who are concerned with radio transmission.

Lastly, and by no means least, I am grateful for the love and support of my wife, Gene, both in reading the whole manuscript more than once despite the unfamiliar content and also for putting up with my long periods spent at the computer.

Ron Kitchen,
Chelmsford,
February 1993

1
Introduction to RF radiation

Radio frequency (RF) radiation

The subject of RF radiation is often regarded as mysterious and something of a black art. This is no doubt due to the fact that it cannot be seen or touched. There was also an element of magic in some of the very early experimental work, particularly that of Tesla, who seems to have mixed science and showmanship.

Perhaps because RF is unseen, it has also become confused with ionising radiation in the minds of many people. It is essential to distinguish the difference between the two since, with our present state of knowledge, the consequences of exposure to them can confidently be stated as being very different.

Although we cannot see radio waves, most people will, at school or college, have done the classical experiments with magnetic fields and iron filings to demonstrate the patterns of the fields and used an electroscope to demonstrate the presence of electrostatic charge and the force which causes the gold leaf to move.

From these early and rudimentary experiments with static fields it should at least be possible to conceive that such fields are not magical and are very common in any electrical environment.

History of radio transmission

Radio transmission is, relatively speaking, a very new technology which had its beginnings in the theoretical work of Clerk-Maxwell in the nineteenth century and the experimental work of Hertz, the German physicist, in the last two decades of the nineteenth century. Many others also made contributions, including the development of devices which could detect the presence of radio waves. Whilst the question of who first transmitted radio signals is not without controversy, the subsequent practical development of radio communications systems is attributed to Guglielmo Marconi who was born in Italy in 1874.

His first British patent was taken out in 1896 and covered the use of a spark transmitter. There are many accounts written of the experimental work carried out at various locations on land and on ships during the course of which the range of such equipment was very much increased. By 1921, the thermionic transmitter tube became available and made it possible to design transmitters to operate on a range of frequencies. The power output available increased with the development of electronic tubes which could, increasingly, handle higher powers with the aid of air or liquid cooling systems.

Over the years, and stimulated by the needs of the first and second world wars, radio transmission has become an established technology which is taken for granted and which, among other things, provides for the broadcasting to our homes of entertainment, news and information of every kind in both the radio and television spheres. The most recent development, resulting in the domestic satellite dish antenna, brings the quasi-optical nature of microwaves to the notice of the consumer.

The use of semiconductor devices (transistors) has become commonplace and as a result the mass and volume of electronic products for a given function is much less than that of their earlier counterparts which used electronic tubes. However, in the high power transmitter field electronic tubes are still the mainstay of transmitters. Semiconductor devices are being used in transmitters of more modest power and also in planar array radar equipments.

Semiconductor devices do have a considerable role in transmitter drives, audio circuits and in control systems. In the latter application, sophisticated logical control circuits are easy to achieve and occupy the smaller volumes attributable to the small size of transistors and integrated circuits.

With the vast increase of terrestrial and satellite broadcasting and communications, homes, work and recreational places are irradiated by a vast number of electromagnetic signals intended to operate receiving equipment, most of which are at very low levels because the high sensitivity of receivers does not necessitate large signals.

Some radiation is unintentional, resulting from the leakage of energy from devices which have no radiation function, for example, due to inadequate shielding, unblocked apertures in metal cases, and similar shortcomings. Apart from any effects of leakage on people, it also causes interference with other equipment. It is not surprising that the presence of so much electromagnetic interference has caused people to question whether they can be harmed by it.

It is worth noting that the original term used for the propagation of such electromagnetic signals was 'wireless' transmission which was descriptive of the communication process in that it implied a change from the established systems of communication over wires (telegraphy and the telephone) to that mysterious arrangement which involved invisible radio waves transmitted over ever increasing distances.

The word wireless has largely passed out of use, being deleted from the author's former company (The Marconi Wireless Telegraph Company) many years ago.

Radio is now the more general term in use. In the domestic field, we have the term radio transmission in use to describe sound broadcasting and the term television transmission to describe television picture and sound broadcasting. There are many words used to describe forms of radio system including satellite communications, radar, microwave links, and personal radiotelephones.

Despite the profusion of terms in use to describe the transmission of intelligence by electromagnetic waves, the nature of these waves is basically the same, the variable being the way in which the intelligence (signal) is added. It is therefore convenient to refer to these electromagnetic waves as 'radio waves' and the frequencies of the waves as 'radio frequencies'.

The nature of radio waves

Most readers will be familiar with the fact that an alternating current or voltage which is undistorted has an amplitude which varies with time and reverses direction at each 180°, one cycle taking 360°. This pictorial representation of a current or voltage is referred to generally as a waveform and the description above is that of a sinewave. Waveforms may have other shapes such as square waves, ramps, etc., as will become apparent later.

A sinewave is illustrated in Figure 1.1 and is shown with the 'Y' axis denoted arbitrarily in amplitude (A). The term amplitude is used to refer to the magnitude of the voltage or current.

The instantaneous amplitude (amplitude at a specified point in time) can be read from such a diagram and will be found to follow a sine curve, i.e., it is equal to the maximum amplitude (A) multiplied by the sine of the corresponding angle.

Hence at 0° and 180° the instantaneous amplitude is zero. Similarly at 90° and 270° the instantaneous amplitude is at the maximum A but since the sine of 270° is negative the polarity and hence the direction of current flow has reversed. This diagram is basically applicable to any simple AC waveform. One of the factors which distinguishes such waveforms is the time duration of one complete cycle (T) in Figure 1.2 and another the frequency (f).

Frequency is simply the number of cycles per unit time and the international convention is 'per second'. The unit is the hertz (Hz) named after the German physicist, one hertz corresponding to one cycle per

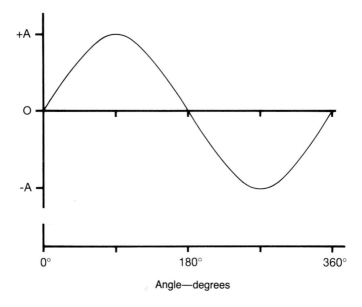

Figure 1.1   *Sine wave illustration*

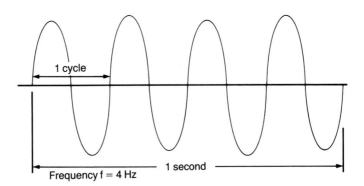

Figure 1.2   *Sine wave frequency and time relationship*

second. It follows that the time of one cycle in seconds is given by the reciprocal of the frequency in hertz.

The AC mains supply frequency (50 or 60 Hz) is referred to as a low frequency whereas the frequencies used for radio transmission are, in the generality, much higher frequencies. The time T for the duration of a cycle at, for example, 50 Hz is 1/50 s = 20 ms (twenty thousandths of a second) whereas the time for higher frequencies is much shorter as shown in the examples below:

T (secs) = 1/f(Hz)

Examples:
1 f = 100 kHz T = 10 μs $(10^{-5}s)$.
2 f = 1 MHz   T = 1 μs $(10^{-6}s)$.
3 f = 1000 MHz T = 1 ns $(10^{-9}s)$.

For those unfamiliar with these SI prefixes (μ, n, etc.), see Table 1.1 which lists those actually used in everyday work.

Table 1.1 *The most commonly used International System (SI) prefixes*

Symbol	name	factor
k	kilo	10^3
M	mega	10^6
G	giga	10^9
T	tera	10^{12}
m	milli	10^{-3}
μ	micro	10^{-6}
n	nano	10^{-9}
p	pico	10^{-12}

If the existence of two identical waveforms as shown in Figure 1.3, is considered, it is possible for these to be displaced along the time axis so that whilst they are identical in form, the starting points of the cycles may not be identical, i.e., there is a phase difference between the two waveforms.

This may be expressed in angular terms, e.g. 45° phase difference. If two such identical waveforms are exactly in phase and are added, the amplitude of the resultant at any point will be twice that of either waveform alone.

Conversely, if the two waveforms are 180° out of phase the sum will be zero. This becomes relevant when considering radiation surveys and it is necessary to consider the additive possibilities of radio waves reflected from the ground and from metal masses. Obviously additions increase any potential hazards whereas cancellations are less significant in this context since the safety measurement activity is essentially concerned with the highest levels present.

Readers will also be familiar with the idea that a current flowing in a conductor gives rise to a magnetic field around it. When such a current is varying, it gives rise to a similarly changing electric field. Similarly a

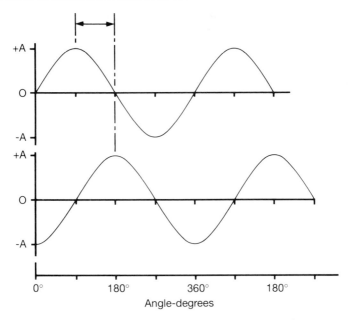

Figure 1.3 *Time phase difference between two waveforms*

changing electric field will give rise to a magnetic field. Unchanging fields of either kind will not result in the production of the other kind of field. With changing fields the magnetic field and electric field are thus inextricably linked. Hence alternating currents and voltages do, by definition, involve time-varying fields.

It is easy to imagine that from any source of such fields some energy may be unintentionally released (transmitted) into free space, causing interference with receivers or other equipment, without necessarily understanding the phenomenon. This is because such 'interference' has been experienced by most people in their everyday lives. Perhaps the most common example is the motor car ignition system which can also prove to be a rudimentary example of the spark transmitter!

In the case of radio transmitters, however, the whole intention is to transmit RF energy into free space and the antenna used to do so is specifically designed to achieve this objective. If we consider the frequencies, discussed above, the very low frequencies, e.g. mains power frequencies, do not give rise to any significant amount of radiation. However, as we increase the frequency then it becomes increasingly possible to radiate electromagnetic waves, given a suitable antenna to act as an efficient 'launcher'.

The electric and magnetic field quantities mentioned above perhaps need a little more elaboration. The electric (E) field at any point is defined as the force acting on a unit positive charge at that point. The magnitude of the electric field is expressed in volts per metre (Vm^{-1}).

The magnetic field at a point is also a force and is defined as the force which would act on an isolated north pole at that point. The ampere is defined on the basis of the magnetic force exerted when a current flows in a conductor and magnetic field strength is measured in amperes per metre (Am^{-1}).

Being forces, both quantities are vector quantities having magnitude and direction. The normal Ohm's law equations for power when the voltage and current are in phase (plane wave conditions) can be used in an analogous way with the same phase qualification to calculate power flux density (usually shortened to 'power density'):

$$S \ (Wm^{-2}) = E \times H = Vm^{-1} \times Am^{-1}$$

Where S = r.m.s. power density; E = r.m.s. electric field value and H = r.m.s. magnetic field value.

Hence S is in units of watts per square metre (Wm^{-2}). In the USA the most common unit used is $mWcm^{-2}$ and is one tenth of the unit expressed in Wm^{-2}, i.e., $1 \ mWcm^{-2} = 10 \ Wm^{-2}$

For plane wave conditions, the ratio $|E|/|H|$ is the impedance of free space (Z_o) and is $377 \, \Omega$.

$$Z_o = Vm^{-1}/Am^{-1}$$

Hence, under the same conditions:

$$S = E^2/Z_o = H^2 \times Z_o$$

Where E and H are the r.m.s. values in Vm^{-1} and Am^{-1} respectively.

Electromagnetic waves propagated in free space have the electric and magnetic fields perpendicular to each other and to the direction of propagation, as represented in Figure 1.4 and are known as transverse electromagnetic waves (TEM waves).

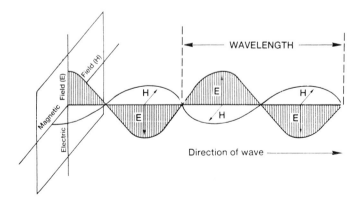

Figure 1.4 *Representation of a plane wave*

The basic nature of an electromagnetic wave can be physically illustrated by holding two pencils with their unsharpened ends touching and the two pencils being mutually at right angles to each other and held so that one is parallel to the ground and one pointing vertically to represent the planes illustrated in Figure 1.4. If now a third pencil is added, mutually at right angles to the other two, it will indicate the direction of propagation as in the figure. The vertical pencil point represents the electric field (vertically polarised wave) and the second pencil the magnetic field.

The plane of polarisation of a wave is, by convention, that of the electric field, i.e., the polarisation in Figure 1.4 is vertical. This convention has the advantage that for vertical polarisation the antenna will also be vertical, e.g. a simple rod antenna and this convention is followed in this book.

If the diagram is rotated until the electric field is horizontal then the wave polarisation illustrated is horizontal. Apart from this 'linear polarisation', other forms such as circular or elliptical polarisation are also used for specific purposes.

Frequency and wavelength

Two related characteristics of electromagnetic waves are used as a method of referencing the waves. They are the frequency (already discussed above) and the wavelength. The latter is denoted by the symbol lambda (λ). The relationship between these two characteristics involves consideration of the velocity of propagation of radio waves.

The velocity of propagation of all electromagnetic waves (c) is constant in a given homogeneous medium and in free space has a value of approximately 3×10^8 metres per second. This figure is also used for air but does not apply to propagation in other media. The relationship between frequency and wavelength is:

$$c = f \lambda$$

Where the wavelength (λ) is the physical length of one cycle of the propagated wave, as shown in Figure 1.4.

For electromagnetic waves in free space, where f is in hertz (Hz):

$$\lambda \text{ (m)} = \frac{3 \times 10^8}{f}$$

Examples:
1   $f = 200 \text{ kHz}$ $\lambda = 1500$ metres
2   $f = 10 \text{ MHz}$ $\lambda = 30$ metres

When f is in MHz, the division simplifies to: λ (m) = 300/f. This lends itself to easy mental arithmetic! Wavelength is an important parameter in considering antenna systems and propagation since it is a factor in determining the physical dimensions of antennas.

Without going into antenna detail at this stage, some idea of the physical comparison of wavelengths can be obtained from the examples of the length dimension of a $\lambda/4$ (one quarter wavelength) antenna for a few frequencies shown in Table 1.2. Practical antennas will be a little shorter than the theoretical calculations of Table 1.2.

Table 1.2   *Nominal quarter wave antenna length for a number of frequencies*

Frequency (MHz)	Length - one quarter wavelength (m/cm)
0.1	750 m
1	75 m
10	7.5 m
100	0.75 m (75 cm)
1000	0.075 m (7.5 cm)
10 000	0.0075 m (0.75 cm)

Radio waves can therefore be referred to either by the wavelength or the frequency. Domestic receivers may have the scaling in either unit but generally frequency is used, as it is in professional radio work. Wavelength does need to be used when it is involved in determining the physical dimensions of antennas and other devices.

In this book, the range of frequency considered is roughly from 10 kHz to 300 GHz. Table 1.3 illustrates the International names for the various

Table 1.3   *International radio frequency designations, bands 4 to 11*

Band no.	Symbols note 1	Frequency range	Metric subdivision
4	VLF	3 to 30 kHz	Myriametric waves
5	LF	30 to 300 kHz	Kilometric waves
6	MF	300 to 3000 kHz	Hectometric waves
7	HF	3 to 30 MHz	Decametric waves
8	VHF	30 to 300 MHz	Metric waves
9	UHF	300 to 3000 MHz	Decimetric waves
10	SHF	3 to 30 GHz	Centimetric waves
11	EHF	30 to 300 GHz	Millimetric waves

Note: E = extra; S = super; U = ultra; V = very.

sub-divisions of the radio spectrum. The term 'microwave' does not appear in the listing although with the advent of microwave ovens it has become widely used in the public domain. There is no generally agreed definition but it is often used to apply to frequencies from something like 300 MHz upwards.

It should be noted that the term 'radio frequency' (RF) is used here across the whole spectrum as a generic term and the term 'microwaves' merely refers to a portion of the RF spectrum.

The abbreviated band identifiers in Table 1.3 from VLF to UHF are in frequent use but the abbreviations SHF and EHF are less used, being now increasingly swallowed up in the loose use of the term 'microwaves'. In .ddition there is a more specific classification system for bands in the upper UHF onwards.

This is given in Table 1.4 on the basis of the 'old' and 'new' listings. It has to be said that different versions of these band classifications are in use across

Table 1.4 *'Microwave' band letters from 1 GHz onwards*

OLD band letter	Frequency (GHz)	NEW (NATO) band letter	Frequency (GHz)
L	1–2	D	1–2
S	2–4	E	2–3
C	4–8	F	3–4
X	9–12.4	G	4–6
J/Ku	12.4–18	H	6–8
K	18–26.5	I	8–10
Q/Ka	26.5–40	J	10–20
U	33–60	K	20–40
V	50–70	L	40–60
E,O	60–90	M	60–140
W	75–110		
T	110–170		

Note: There are many variants of these classifications.

the world and in textbooks so that reference to frequency is perhaps the only safe way of avoiding ambiguities. The presentation of the different possible classifications tends to confuse rather than enlighten.

At the bottom end of the spectrum, the abbreviation ELF (extra low frequency) has been adopted by those working with low frequencies to include mains supply power frequencies.

Conveying intelligence by radio waves

When a wave of a given frequency is radiated continuously, i.e., a continuous series of sine waves, no intelligence is conveyed and the signal is called a 'carrier'. This mode of transmission is known as continuous wave (CW).

If the carrier is switched on and off in accordance with some kind of code, e.g. morse code, then this intelligence can be interpreted. More generally, for broadcasting, the intelligence may be speech, music and television pictures. Other professional work includes voice and data transmission by a variety of methods, radar transmitters transmit RF signals in a series of pulses and so on.

The process of sending intelligence is referred to as the process of modulation and the technical methods of doing so are wide ranging and outside the scope of this book. It is however useful to illustrate the general nature of amplitude modulation which has some significance when carrying out radiation measurements, and also to illustrate the principle of pulse transmission.

Figure 1.5 illustrates the waveform of a carrier and of the same carrier with 50% modulation applied in the form of a simple audio frequency sinewave. It can be seen that the instantaneous amplitude of the 50% modulated wave is 1.5 A against A for the carrier. Clearly the total power is greater when the carrier is modulated and hence any field measurements made will need to be related to the modulation state. For amplitude modulation, Figure 1.6 shows the relationship between sine wave amplitude modulation depth versus transmitted RF power and RF current.

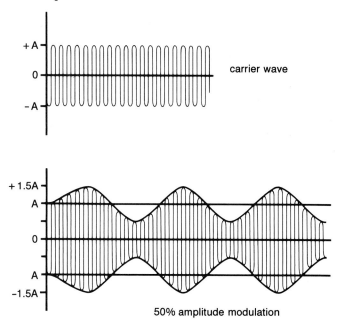

Figure 1.5 *RF signal, amplitude modulated by a low frequency signal*

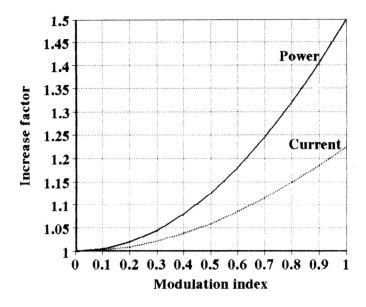

Figure 1.6 *Sine wave amplitude modulation depth versus RF power and RF current*

Figure 1.7 shows a pulse transmission where the carrier is transmitted for time tp (the pulse duration) and with a pulse repetition rate of n Hz (pulses per second). It is of course, not possible to show this to scale since there will be too many cycles of carrier in each pulse to actually illustrate them. For example a radar working at 1 GHz and with 2 μs pulses will have 2000 cycles of carrier in each pulse.

There are many other methods of modulation and transmission which can be applied to radio equipment and which cannot be covered here but which need to be known to those doing safety surveys.

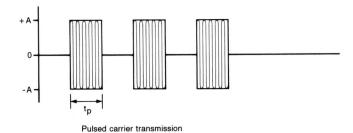

Pulsed carrier transmission

Figure 1.7 *Pulse modulation*

Ionising and non-ionising radiations

Confusion between these two forms of radiation amongst the public has been mentioned earlier. From the author's own experience, there is also a surprising amount of misunderstanding amongst electronics and radio engineers about the distinction between these two forms of radiation so that RF radiation is sometimes considered to be the same as ionising radiation.

Ionising radiation, by definition, is radiation capable of ejecting electrons from atoms and molecules. There is a minimum quantum energy below which this disruption cannot take place. Since the human body is largely water, the water molecule is used to define this minimum level.

Different reference sources give varying figures for this between 12 electron volts (eV) and 35 eV. The actual value does not matter for the purposes of this comparison. 12 eV corresponds to a wavelength of 1.03×10^{-7} metres which can be seen from Figure 1.8 lies in the ultraviolet spectrum.

The highest RF frequency used in standards for RF safety is 300 GHz which corresponds to a wavelength of 10^{-3} metres and lies in the EHF band of the radio frequency spectrum. If the calculation is done the other way round, 300 GHz corresponds to an energy of 0.00125 eV which, from the foregoing, is far too small to cause ionisation.

However, in radio transmitters using very high supply voltages, ionising radiation in the form of X-rays are produced and for this reason the subject is covered in Chapter 7. It should be clear that this ionising radiation is not inherent in the RF energy but rather that both forms of radiation can co-exist inside equipment and the RF engineer or technician needs to be aware of the hazards involved. It is also the case that ionising radiation is, in most countries, subject to definitive legal provisions due to its hazardous nature.

Explanation of terms used

In this section those terms and units which are most frequently used in dealing with RF radiation are explained. The more formal definitions may be found in reference books. Other more specialised terms are introduced in the text as appropriate. Abbreviations are given in Appendix 1. Common units and conversions are given in Appendix 2.

1 Transverse electromagnetic mode wave (TEM)

An electromagnetic wave in which the electric and magnetic fields are both perpendicular to each other and to the direction of propagation (see Figure 1.4).

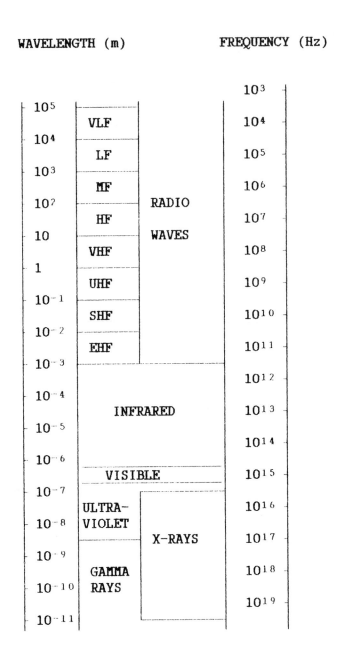

Figure 1.8 *Chart of frequency versus wavelength across the spectrum*

2 Power

The rate of doing work in joules per second. The unit is the watt (W) which corresponds to $1 \, \text{Js}^{-1}$. Sources of RF energy are rated in watts. Both kilowatts and megawatts are common in radio work, the latter typically for very high power equipment such as radar equipment.

3 Mean power

The power supplied or generated averaged over a period of time which is long compared with the longest period of any modulation component.

4 Power flux density (power density)

Power flow per unit area through a surface normal to the direction of that flow, usually expressed in watts per square metre (Wm^{-2}). However it is also often quoted in mWcm^{-2}.

The use of hybrids such as Wcm^{-2} are best avoided since they can cause confusion. In this book the shorter form in common use, 'power density', is used hereafter. All references to power density, electric field and magnetic field are to r.m.s. values, unless otherwise stated, in common with the practice in RF safety standards.

5 Energy density

This is, strictly, related to volume (Jm^{-3}) but is almost universally used in radiation protection work as the product of power density and time and expressed either in watt-hours per square metre (Whm^{-2}) or joules per square metre (Jm^{-2}). $1 \, \text{J} = 1 \, \text{Ws}$. It is sometimes used to express a total energy limit, for example, 'not more than $5 \, \text{Whm}^{-2}$ in a six minute period'.

In terms of the energy in a volume, e.g. Jcm^{3}, the definition relates to the energy in a minute volume divided by that volume. With a power density of $10 \, \text{Wm}^{-2}$ the energy in a cubic centimetre of air is 0.033 picojoules.

6 Electric field strength (E) at a point

A vector quantity defined as the force acting on a unit positive charge at that point. It is expressed in volts per metre.

7 Magnetic field strength (H) at a point

A vector quantity defined as the force acting on an isolated north pole at that point. It is expressed in amperes per metre.

8 Specific absorption rate (SAR)

The rate of absorption of RF energy in a substance, normally human tissue, expressed in watts per unit mass, e.g. watts per kilogram. If the substance is not human tissue, it should be specified. Note that an SAR limit may be expressed in this standard form but be limited to a maximum mass of tissue, e.g. $10\,\mathrm{Wkg}^{-1}$ ($10\,\mathrm{g}$) should be interpreted as an SAR of $10\,\mathrm{Wkg}^{-1}$ in any 10 grams of tissue.

9 Frequency

The number of cycles of an alternating current per unit time where the International period is one second. The unit is the hertz. $1\,\mathrm{Hz} = 1$ cycle per second.

10 Pulse repetition frequency (p.r.f.)

In a system which uses recurrent pulses, the number of pulses occurring per unit time. The unit is the hertz (Hz).

11 Peak pulse power density

In pulsed systems such as radar equipment the term peak pulse power is used when what is actually meant is the mean power in the pulse (see Figure 1.7). This should not be confused with instantaneous peak power.

12 Pulse duty factor (DF)

Where t_p is the pulse duration in seconds and n is the pulse repetition rate in Hz, then the duty factor $DF = t_p n$ and has a value less than 1.

For example, if $t_p = 2\,\mu s$ and $n = 500\,\mathrm{Hz}$, then:

$$DF = 500 \times 2 \times 10^{-6} = 0.001.$$

Many people find it easier to work with the reciprocal, in this case $1/0.001 = 1000$.

The relationship between peak pulse power density (S_{pk}) and the mean power density (S_{mean}) in a pulsed system is:

$$S_{pk} = S_{mean}/DF \text{ or, if using the reciprocal of DF:}$$

$$S_{mean} \times 1/DF = S_{pk}$$

Note that although pulse transmission often seems to be uniquely linked to 'radar', pulse transmission is widely used and radar is just one application. Note also the high values of S_{pk} which are possible, depending on the duty factor.

13 Antenna (aerial)

The generally used term for any type of device intended to radiate or receive RF energy. These range from simple wires and rods to arrays (of which the television antenna is an example) to large microwave parabolic, elliptical and rectangular aperture systems.

Some antennas are dedicated to reception or transmission whilst others do both. To most people the terms antenna and aerial are synonymous. The English plural is normally used for antenna.

14 Antenna, isotropic

A hypothetical, idealised antenna which radiates (or receives) equally in all directions. The isotropic antenna is not realisable but is a valuable concept for comparison purposes.

15 Directive gain of an antenna

The ratio of the field strength at a point in the direction of maximum radiation to that which would be obtained at the same point from an isotropic antenna, both antennas radiating the same total power.

16 Antenna beamwidth

The angular width of the major lobe of the antenna radiation pattern in a specified plane. The usual criterion for beamwidth is to measure between points either side of the beam axis where the power density has fallen to half (3 dB down) of that on the axis. This is usually referred to as the '3 dB beamwidth'.

17 Equivalent radiated power (ERP)

The product of the radiated power and the gain. Often used to specify the power of UHF/VHF broadcast transmitters.

18 RF machines and RF plant

RF energy is now increasingly used to undertake manufacturing operations which use heating and these terms are used here to refer to machines generally. In practice they have functional names e.g., plastic bag sealer, plastic welder, etc. Their significance is that they use an RF generator which, in terms of safety, needs the same consideration as any other RF generator.

Use of the decibel

Whilst most people trained in electrical and electronic engineering will have covered this topic, experience in running RF radiation safety courses shows that whilst many people work regularly with decibels, a surprising number never have occasion to do so and it is often necessary to do a refresher session in this topic.

The bel and the decibel (one tenth of a bel) was originally used to compare sound intensities and is currently used in safety legislation to limit the exposure of people to intense sounds in the workplace. Some safety officers will be familiar with this method of noise control.

In radio work, the decibel is used to compare powers, voltages and currents. The decibel is a dimensionless number representing a ratio based on common logarithms. However, usage is such that the ratio is often referenced to a value of a quantity so that it can be converted to a specific value of that quantity. This is a practice of convenience which has developed, so it is best to start with the basic role of the decibel as a dimensionless number. The bel itself is not normally used in radio work.

Decibels and power

If we wish to compare two powers, P_1 and P_2, then we can do so by dividing one by the other. The resulting ratio is P_1/P_2 and is a pure number. To express this in decibels the form is:

Ratio (dB) = $10 \log(P_1/P_2)$

If $P_1 = 1600\,W$ and $P_2 = 2\,W$ then the simple ratio is:

800

The ratio in decibels is $10 \log 800 = 29.03\,dB$

Since the decibel is based on logarithms, a number of simplifications follow. The basic rules for ratios which are pure numbers are therefore:

1 Multiplying numbers merely requires the addition of the decibel values.
2 Dividing numbers requires the subtraction of one decibel value from the other.

The basis of the first part of Chapter 5 is to use dB ratios so that only simple addition and subtraction is needed. As powers can be in kilowatts or megawatts, it can be seen that the arithmetic involved is much simpler, especially as gains can also involve inconveniently large numbers, e.g. 69 dB gain = 7 943 282.

To convert decibel values back to plain ratios we reverse the process:

For 29 dB, the ratio is given by:

Antilog (29.03/10) = 800 as in the first calculation.

Decibels and voltage

Since power can be expressed as V^2/R, then the ratio of two such expressions where, V_1 and V_2 are the two voltages and R_1 and R_2 the corresponding resistances, is:

$(V_1^2 R_2)/V_2^2 R_1$

If $R_1 = R_2$, then the ratio is now:

$(V_1)^2/(V_2)^2$

and dB $= 10 \log (V_1)^2/(V_2)^2$

$= 20 \log V_1/V_2$

Hence for voltage ratios, the formula for conversion is:

Voltage ratio (dB) $= 20 \log$ voltage ratio.

Referencing ratios

So far we have considered dimensionless quantities where the rules for handling the resultant dB values are those related to the use of logarithms generally. It is possible to reference ratios to any quantity, a common one being the milliwatt. The usual reference to this is dBm rather than the expected dBmW. In Chapter 5 some calculations are referenced to 1 watt ($dBWm^{-2}$). Table 1.5 shows some decibel values referenced to 1 watt for powers greater and smaller than the reference value. This is a very convenient way of handling power in calculations.

Table 1.5 *Table of power (watts) versus decibel value relative to 1 watt*

Watts	dBW
1000	+30
100	+20
10	+10
1	0
0.1	−10
0.01	−20
0.001	−30

To convert back to watts power, the process is as before except that the ratio obtained is multiplied by the reference value:

1000 W = 30 dBW

To reverse this:

Power (W) = antilog (30/10) × reference value 1 W
            = 1000 × 1 (watts)

When the reference value is unity, in this case 1 watt, the last multiplication is academic.

2
Sources of RF radiation

Introduction

RF equipment is now extremely widely used in applications which would not have been conceived twenty or thirty years ago. Apart from the enormous diversity of equipment available in the established fields of communications, broadcasting, radar, navigation, production processing and medical therapy, there is an increasing use in applications such as anti-theft systems in shops, vehicle location, motorway control, telemetry to operate control systems remotely and many other novel applications. RF has even been used experimentally to break up concrete.

This chapter aims to give an outline of the sort of RF sources which may concern those responsible for RF radiation safety, together with some illustrations of the sort of equipment which might be met.

Broadcasting

MF and HF broadcasting

Broadcast transmitters in the MF and HF bands use considerable power, 300 kW and 500 kW being common.

An article by Wood discussing international broadcasting in the Arab world [1] showed transmitter sales to this area in terms of system power and listed nine long wave and medium wave stations with powers exceeding 1 MW, most of them being rated at 2 MW. In the HF broadcast field one station was shown as having 16 × 500 kW transmitters. The potential safety hazards associated with the feeders and antenna systems can be imagined.

MF transmitters are usually used for national broadcasting to give wide coverage whilst HF equipment is used for long distance broadcasting, for example, as used by the BBC overseas service and the Voice of America broadcasting service.

By their nature and size, high power HF broadcast transmitters provide a good example of equipment which requires quite a lot of survey time because of the need to measure RF and X-ray radiation safety on a number of different frequencies. Figure 2.1 shows a 500 kW HF broadcasting

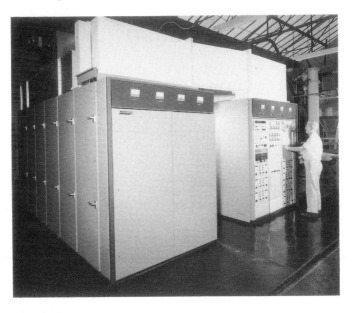

Figure 2.1 *500 kW HF broadcast transmitter*
(Courtesy Marconi Communication Systems Ltd)

transmitter. It consists of two cabinet structures, one for the transmitter and the other for the modulator.

The frequency range is 3.9 to 26.1 MHz and the types of modulation include amplitude double sideband modulation and single sideband with reduced carrier. The modulation system for double sideband amplitude modulation is the Marconi patent 'pulsam' high efficiency pulse modulation system. The transmitter has some novel features such as copper tube inductors with cooling water flowing through them and the associated vacuum capacitors.

By virtue of the physical size of the equipment, this transmitter illustrates the amount of RF and X-ray leakage testing needed on large equipment. There is something like 70 square metres of panel surface area to survey for leakage, about half of this being the 'active' surface area where leakage is possible.

With an output power of 500 kW, the output voltage and current will be very high into a 50 Ω load, or into the 75 Ω or 300 Ω alternatives.

Antenna systems used for MF may be wire systems or towers fed directly so that they become the radiators. For HF broadcasting wire curtain arrays, rhombic and other types may be used. These involve a lot of masts and towers to support the arrays. The feeders are often 300 Ω open wire types or 50/75 Ω coaxial types. Since more than one frequency may be used over a twenty-four hour period, additional antenna systems are required. Unused antenna systems can become 'live' due to parasitic energisation from working antennas.

Figure 2.2 illustrates a particular installation of HF wire arrays to give some impression of the safety assessment problems associated with such systems. It can be seen that the job of ensuring the safety of riggers and maintenance personnel is not an easy one.

Figure 2.2   *HF antenna arrays*
(Courtesy Marconi Communication Systems Ltd)

UHF and VHF broadcasting

Television and VHF radio broadcasting is now taken for granted in most of the world. The number of broadcasting stations has increased considerably over the last thirty years and the need for full coverage with television and VHF radio has resulted in many lower power repeater transmitters being used to bring the services to local communities.

High power antennas for television broadcasting are usually situated on towers. The frequencies used range from 470 MHz to 850 MHz. The antennas generally use arrays of panels typically using a row of full wavelength slots fed by transmission lines against a conducting back shield.

Either the complete antenna is enclosed in a weatherproof cylinder or the individual panels have their own 'radome' coverings. Figure 2.3 illustrates the general appearance.

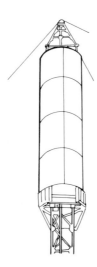

Figure 2.3   *General appearance of a UHF antenna system*

The antenna systems are normally powered in two halves (upper and lower) which improves service reliability and allows maintenance on one half whilst the other half is run on reduced power or shut down. Where the individual tiers of the two halves are interleaved, as is sometimes the case, this does not apply.

In the United Kingdom, the towers used to carry high power antennas are generally very high ones and located away from structures where people might become exposed. Radiation is generally omnidirectional and the antenna array is so arranged as to direct the far field radiation at or just below the horizon. For this reason hazards will generally be related to the tower structure around the antenna and people at ground level should not experience any significant value of power density. The equivalent radiated power (ERP) ranges from 1 MW downwards. Local television repeater stations can have very low ERPs.

High power VHF services usually use tiers of panels such as that shown in Figure 2.4, arranged as for the UHF antennas. As with the UHF systems the VHF systems are often operated in two halves. Although interleaving is not used with VHF, the unpowered half of such an antenna is likely to be driven parasitically causing hazardous areas near the apparently unused antenna. High power VHF antennas are often mounted on the same tower as the television UHF antennas.

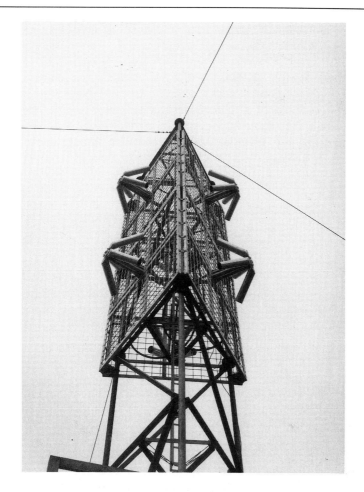

Figure 2.4 *VHF broadcast antenna element*
(Courtesy Marconi Communication Systems Ltd)

In places where high power UHF and VHF broadcasting antenna systems are located on buildings in residential areas, the problems may be more evident than where they are located away from buildings.

Communications

There is an infinite variety of communications equipment ranging from the familiar hand-held mobile two-way radio, which can use HF, VHF or UHF, through MF and HF systems for ground, air and ship communications to microwave systems for terrestrial and satellite communications. In the more domestic environment citizens band and amateur radio transmitters are used, as are radio telephones.

Tropospheric scatter systems

One useful example of a microwave system is the tactical tropospheric scatter system illustrated in Figure 2.5. Tropospheric systems for land communications are trans-horizon microwave communication systems.

Figure 2.5 *Tactrop-transportable trans-horizon telecommunication system*
(Courtesy Marconi Communication Systems Ltd)

Long hop distances (up to several hundred kilometres) are obtained by deflecting high power microwave signals off the tropospheric layer of the atmosphere to overcome the earth's curvature between widely separated sites. Systems may have antennas varying in size from 3 metres to 27 metres, the very large antennas being used in fixed systems.

The tactical system illustrated here is mobile with both military and civil applications. It comprises two antennas of 6.2 metres diameter, an equipment shelter containing all the electronic equipment and a container with dual diesel generators and fuel tank.

The transmitter power from the two power amplifiers is adjustable up to 1 kW. The frequency band is 4.4 to 5 GHz on this equipment. The system is a digital one with 60 encrypted telephone channels available.

In order to improve the path reliability, various combinations of quadruple (four path) diversity operation are available. 'Diversity' operation here involves the simultaneous use of different frequencies, antenna spatial positions or polarisation to receive on four independent paths, combining the outputs to minimise fading effects on reception. There is an alternative antenna option which uses only a single antenna with a dual angle feedhorn giving angle/frequency four path diversity.

With 1 kW on the antennas, which are relatively near the ground, safety surveys are clearly important. The general approach is to have a prohibited zone in front of the antennas.

Air traffic control (ATC) communications

Air traffic control communications systems involve highly organised networks of VHF and UHF transmitters and receivers to communicate with aircraft in the ATC control zone. Usually several transmitters are operated simultaneously to secure adequate zone coverage.

Figure 2.6 shows a typical tower installation of UHF and VHF antennas on the near tower and the far tower. The typical transmitter power output is about 50 watts. The antennas are spaced away from the tower with horizontal booms which also reduce the exposure to those doing work on the tower. The tower aspect is discussed in Chapter 9. Note that many ATC sites also include radar systems which may irradiate parts of the towers and which may need to be taken into account.

Satellite communication systems

Satellite communication systems use microwave beams to communicate with satellites. The diameter of the dish antennas can vary from a few metres to tens of metres and may have very high gains. The narrow beams are used at suitable elevations for the appropriate satellite and give rise to very little radiation exposure at ground level away from the antenna.

Figure 2.6 *ATC VHF and UHF communications antennas (Courtesy of the Civil Aviation Authority)*

Figure 2.7 shows a typical satellite dish mounting in use with Mercury Communications Ltd. The azimuth and elevation settings vary according to the satellite in use. Sometimes problems arise with building workers doing work in the vicinity above ground because of the fact that the antenna may seem to be pointing at them. In fact, the setting of antennas has to avoid nearby buildings due to the attenuation introduced if the beam is intercepted by buildings ('beam blocking').

Figure 2.7 *Satellite communication ground station antenna (Courtesy Mercury Communications Ltd, London)*

Radar systems

There are many varieties of radar equipment in use around the world. Most involve movement of the antenna system, i.e., rotation or movement in azimuth, movement in elevation, etc. Leaving aside HF radar, radar systems are generally characterised by using microwave beams which are usually relatively narrow in azimuth but the characteristic in the elevation plane depends on the nature and function of the radar.

The applications include:

1 Defence
2 Air traffic control
3 Meteorology and the study of weather changes
4 Mapping the earth
5 Specialised applications ranging from radars for measuring the state of the sea and sea wave motion, to hand held police radar speed meters for checking motor vehicle speeds.

They may be ground based (fixed or mobile), shipborne or airborne (aircraft and satellites).

Ground radar systems may be deployed 'naked' or housed in weatherproof radomes (non-metallic domes transparent to radar) which can protect both the equipment and the personnel from exposure to severe weather. Figure 2.8 illustrates a typical radome used on a civil aviation site.

Figure 2.8 *ATC protective radome for a surveillance radar (Courtesy of the Civil Aviation Authority)*

In this case the height keeps the beam away from personnel below. However, this same factor makes beam surveys more difficult due to the height! Sometimes rising ground nearby, as in the left of this picture, makes it possible to obtain access to the beam for survey purposes. By the same token it may irradiate the higher ground, depending on the beam elevation. If people have access to the land this factor needs consideration during a survey.

Figure 2.9 illustrates a CAA surveillance radar (the large antenna) mounted on the top of a concrete tower. Surveillance radars of this type rotate continuously (typically at 6 rpm) to obtain the azimuth information on targets. The smaller antenna on top of the large antenna is a secondary surveillance radar [73] for target identification 'friend or foe' (IFF). This uses much less power than the main radar. Here again, the height of the radar is helpful in avoiding significant personnel exposure at ground level.

The other antennas on the tower are for microwave links.

In contrast, Figure 2.10 illustrates a mobile radar of the planar array type. The mean power is 3.6 kW and the effective peak power 53.5 kW and operates on 23 cm wavelength (nominally 1300 GHz). The antenna rotates on the trailer mount, the speed being 6 rpm. The receive beams are electronically switched. Since both the azimuth and elevation information is provided by the processing, no separate height finder is necessary.

An SSR antenna for IFF is provided and can be seen fitted above the main antenna. From the survey point of view the system is easy to deal with and can be treated as any other conventional surveillance radar. The system is safe to walk around at ground level due to the mount height. The antenna

Figure 2.9 *ATC surveillance and SSR antennas on a tower (Courtesy of the Civil Aviation Authority)*

has very low sidelobes such that in most situations they will not be of any consequence. The electronic processing equipment is fitted in cabins which can be located alongside.

Some large modern military planar array radars have no mechanical movement, all the functions being done by electronic beam switching.

A very different type of radar antenna is shown in Figure 2.11. This is a naval surveillance radar and is shown fitted to the mast of a naval ship. The radome mounted antenna system comprises an air search antenna mounted back to back with a sea and low air search antenna to give a compact medium to short range defence cover.

The antenna assembly rotates at 30 rpm. The picture illustrates the complexity of ship installations, as many other antennas can be seen situated on the structure.

Figure 2.10 *Martello long range 3D surveillance radar
(Courtesy Marconi Radar Systems Ltd)*

A further ship system for the defence field is shown in Figure 2.12 which illustrates the Seawolf radar. This tracks both the target and the missile. It is capable of controlling both guns and missiles. The tracker antenna assembly also carries duplicate command links to transmit missile position and guidance information. Trackers use a very narrow and intense beam, i.e., it has a high power density close to the antenna.

A more well-known type of radar is the approach control system for airfield control illustrated in Figure 2.13. This is a radar with a mean power of 550 W and a peak power of 650 kW. The frequency coverage is 2.7 to 2.9 GHz. It uses a high efficiency two beam antenna system and transmitters in dual diversity. The mounting height minimises problems with people on the ground. Close access on the building roof needs controlling, but access is necessary for maintenance.

Mobile radars of the conventional type usually consist of the antenna assembly, one or more equipment cabins and a power generator. The

Figure 2.11   *Naval surveillance radar*
(Courtesy Marconi Radar Systems Ltd)

transmitter output power is coupled via a waveguide run to the antenna. Part of the survey activity will therefore include leakage testing along the waveguide run. Most fixed radars will also have such waveguide runs which need careful leakage testing.

Aircraft, both civil and military, carry powerful radar equipments. Problems may arise if such equipment is left running on the ground or if ground tests are not conducted under suitably controlled conditions. Similar problems can arise with aircraft communications, navigational and military jamming equipments where control is needed for ground testing and servicing.

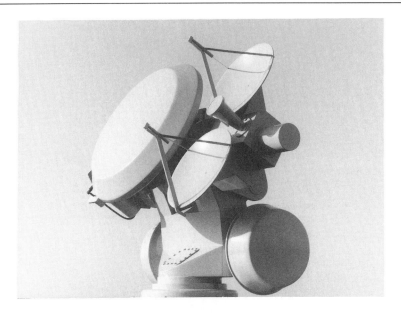

Figure 2.12 *Seawolf tracking radar*
(Courtesy Marconi Radar Systems Ltd)

Figure 2.13 *Airfield approach control radar on the roof of an airfield control tower*
(Courtesy Marconi Radar Systems Ltd)

RF machines

The use of RF generators in manufacturing processing is very widespread, covering a large range of materials and processes. Low frequency induction heating has been used for at least 50 years for metal processes.

Many of us in communications, radar and electronics research and development generally may not be very familiar with modern RF processing machines which are now being used widely in industries which are not classified as part of the electronics industry. There are of course such machines in the electronics industry for bag sealing, packaging and similar applications as well as many specialised equipments in electronics research.

However it it is probably true to say that RF machine use is growing much more rapidly in basic processing industries where flow-line operations can be used on large volumes of products. The illustrations which follow show some of these developments.

Processing machines

The invention and world-wide marketing of the microwave oven, domestic and industrial, heralded a widespread use of RF generation in the food processing industry. RF drying systems are now used as standard practice to improve the quality and colour of food products such as biscuits (cookies), bread products, crispbreads, sponge products, cereals and snack foods.

Such equipment facilitates flow-line production with the product being fed to the RF applicator via a conveyor system. Figure 2.14 shows a

Figure 2.14 *A Strayfield 100 kW 27 MHz post-baking installation for volume production of rusk-type products (Courtesy Strayfield International Ltd)*

Strayfield 100 kW 27.12 MHz post-baking unit used on rusk-type products. The product is transported through the equipment on a conveyor system so that the operator does not have to have close access to the RF source.

It will be noted that the size of the installation is comparable with a fairly large radio transmitter. A large production organisation may have a range of such installations processing a variety of products so that the whole concept of a food production system is changed to one biased towards electronic equipment. Indeed the first quick glance at a photograph of such an organisation gives the impression of a large transmitting station. In the literature of this manufacturer, many well known brands of foodstuffs can be seen emerging from RF machines!

RF drying equipment is also used for drying textile materials, drying materials which have been dyed, the paper processing industry, woodworking, drying lacquer and other coatings and similar applications.

Industrial and domestic microwave ovens generally work on 2450 MHz and are in very wide use for cooking food. With the relatively shallow penetration of RF at this frequency, the cooking of meat joints and similar large objects depends on conduction of heat from the heated outer layers of the meat to the inner layers. Microwave heating has other applications in the medical field for example, the rapid thawing of sachets of frozen blood.

Other applications of RF process machines include plastics processing and welding, vulcanising, heat sealing and packaging, crystal growing, brazing, soldering, hot forging and many others. Figure 2.15 shows an induction

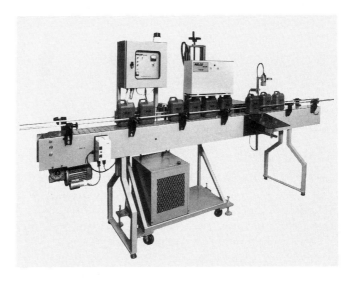

Figure 2.15   *A Relco 2 kW 27 MHz RF induction heater in use on volume production sealing oil containers*
(Courtesy Relco UK Ltd)

heating machine sealing oil containers on a flow-line. The generator is rated at 2 kW at 27 MHz. The containers are sealed at a rate exceeding 200 containers per minute. The complete system can include a detector to identify any completed products lacking a sealing foil. Another detector can sense whether the incoming flow of products is present. The RF output can be interlocked with conveyor movement.

Container sealing is used for a number of reasons. For food products the aim may be to exclude oxygen. It may also be to provide tamper evidence in the same way as pharmaceutical products which routinely use methods which highlight tampering. For agricultural chemicals and other substances the main reason may be to reduce the possibility of leakage.

Figure 2.16 shows a Stanelco TP3/15 machine rated at 15 kW 27.12 MHz for plastic welding. The products are fed by means of a double shuttle tray.

Figure 2.16   *Stanelco TP3/15 kW Mk 2 plastics welder with double shuttle tray for loading*
(Courtesy Stanelco Products Ltd)

The machine is fitted with fully interlocked guards for operator safety. The list of suitable applications is again large. Examples include: inflatables, upholstery, medical products, baby pants, wallets and blister packing.

Discussion with representatives of the manufacturers whose products have been illustrated above indicated that they are well aware of the hazards of RF radiation and the need for adequate shielding on RF machines. Some of them were involved in the drafting of the UK guidance note PM51 [50] referred to later. There is also a UK committee to which most RF machine manufacturers belong, the British National Committee on Electroheat (BNCE).

Medical therapy equipment

Belief in the beneficial effects of magnetic fields and electrical currents applied to the human body dates back to the late 19th century. 'Short wave' (HF) therapy has been used in hospitals and clinics since the early twentieth century. Some of the very early claims were somewhat dubious but serious medical people have devoted their time in trying to explore the benefits and devise effective therapies.

The equipment used was usually a 27 MHz equipment since this frequency gives a significant penetration in the human tissues and this is necessary for those therapies which essentially aim to induce heating in joints. According to a challenging paper by Barker [2] too little work has been done to scientifically test the benefit of RF therapies by recognised 'double blind' test methods.

Nevertheless there are large numbers of RF therapy machines in use in hospitals and a wide variety of new designs in the suppliers' catalogues. The machines used seem to make most use of pulsed RF with various permutations of pulse width and repetition rate available to the physiotherapist.

Work done in the 1970s in the USA seems to have established a high success rate with the use of a pulsed RF field applied by coils to bone fractures, particularly those which were slow to heal.

An 80% success rate on 30,000 patients is recorded by Barker, though he does point out that it has not been scientifically demonstrated that the RF field has added to the incidental immobilisation inherent in the treatment regime.

More recently microwave frequencies have been used though it does not seem clear what benefits are considered to result. One exception may be the use of microwave equipment to induce localised hyperthermia to destroy malignant tumours (Chapter 3). From the point of view of safety surveys, we are concerned with:

1 Exposure of the patient – basically a medical matter.
2 Exposure of the attending physiotherapist, which is a general health and safety matter.
3 Any exposure of other people nearby.

Whatever the frequency and the equipment used, the safety measures will include careful control of the patient's exposure to the wanted radiation, and keeping the patient at a reasonable distance from stray radiation from the equipment applicator leads. For the physiotherapist there is a need to minimise exposure since there can be high field levels near the equipment and its leads.

This is best arranged, wherever possible, by keeping away from the equipment except when setting it. Some equipment includes the provision of a switch for the patient so that if there is discomfort, the patient can switch it off. Third parties ought not to be exposed unless it is medically necessary, so the separation of equipment and people is important.

Figure 2.17 shows a typical modern equipment, the EMS Megapulse Senior, which can supply up to 375 watts at 27.12 MHz. Energy is applied via capacitor electrodes, rigid and flexible. Pulse repetition rate is adjustable and the pulse width can be set between 20 and 400 μs. Continuous RF as well as pulsed RF can be used. Such equipments have to be tuned with the applicators applied to the patient. Some require manual tuning and others offer automatic tuning. This equipment is of the latter type.

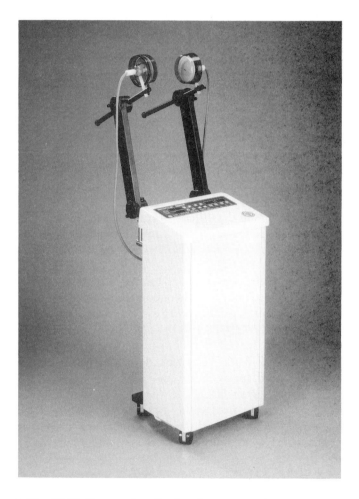

Figure 2.17 *EMS Megapulse Senior therapy equipment (Courtesy of Electro Medical Supplies)*

General

A reasonable basic knowledge of equipment is needed to carry out surveys since otherwise some safety aspects may be overlooked. The nature of the RF source, in terms of the power rating, types of modulation used, pulse characteristics and duty factor, frequencies used, etc., needs to be known in order to determine the worst case situation for RF radiation.

Some modulation techniques may not be familiar to the surveyor and advice may need to be sought.

The HF band is, from the survey point of view, particularly sensitive to frequency since the general characteristic of all RF safety standards is a tightening of safety limits over this band so that the permitted limit at the top of the HF band is much lower than at the bottom end.

This aspect is relevant to transmitters and to many RF process machines. Modern transmitters, particularly military systems, may have frequency hopping facilities (where the transmitted frequency changes periodically for security purposes) and other peculiarities which may have implications for survey work.

There are many other sources of RF power including signal generators for testing work which can offer tens and hundreds of watts output. Often the magnitude of the output of such devices is not appreciated by the user. This is particularly the case where generators are used to power microwave waveguide benches.

With RF processing machines, some knowledge of the effect of different processing procedures and the different types of work pieces involved may be needed for machines which process a variety of different products, since the leakage and possible operator exposure may be affected by the type of product.

Cases have been reported where operators feel a sensation of heat in their hands when loading machines and where operators sometimes find that their footwear gets hot due the to heating of eyelets, rivets and similar parts in boots. There is therefore a good reason to discuss with machine operators any effects which they have noticed.

Shielding around the work piece may, on manually operated machines, conflict with the need for visibility and may require resort to modern transparent shielding materials.

In summary, there is a need to be aware of all equipment which may generate significant RF power and the nature of the operation of each item. The maintenance of a register of such items will usually be a sensible provision.

3
Effects of RF radiation on people

The nature of potential hazards

A great deal has been written over the last thirty years or more about the hazards of RF radiation. The vast majority has been in the form of serious contributions and includes a large number of research papers. Lack of accurate methods of measuring fields obviously affected some of the work of the earliest workers. As technology has improved and field measurements can now be made more accurately, the experimental methods have improved.

The problems of research in this field are fairly obvious – very few tests can be carried out on human beings. As a result, most practical work has been done on small animals such as rats, mice, rabbits, bacteria, yeast cells, fruit flies and similar subjects. There is then the problem of the extrapolation of the results to human beings when some fundamental factors, e.g. the physical sizes, and thus the resonant frequencies of the various subjects are so markedly different. Also, where thermal effects are involved, the differences in the thermo-regulatory systems of the test subjects pose a very considerable problem. Hence such extrapolation is likely to be dangerous.

An even greater problem is the fact that the radio frequency spectrum is so very wide (perhaps 10 kHz to 300 GHz) and it is well nigh impossible to extend research to the whole spectrum, to low and high levels of field, different modulation methods and so on. This is further complicated by the suggestion that some effects only occur in RF frequency 'windows' and modulation frequency or pulse rate 'windows'.

The term 'window' here implies that an effect has been claimed to occur at some RF frequencies and not at others or at low field levels and not at higher field levels of the same frequency or that the effect occurs at certain modulation rates and not at others. The variables, RF frequency, RF amplitude, modulation frequency and type, provide an almost infinite number of combinations to be studied. This illustrates the real difficulty in determining which combinations to explore. In more practical terms there is also the cost of equipment capable of generating all the frequencies and modulations at levels large enough for practical work.

It is scarcely surprising that from time to time some particular research may be challenged, either because of something related to the experimental situation or because of the conclusions drawn. In general the replication of research findings elsewhere is looked for but is not easy to achieve when finance is not available to pay for the work.

Some individuals express extreme views on RF radiation hazards, which are often drawn from the research of others, but which differ from those of the researchers concerned and from those of others working in the field of RF radiation. Where such views are genuinely held and the person concerned has a reasonable competence to handle the research concerned, this is no problem. It may even provide some impetus for more research, if the views are not so extreme as to be ignored.

The real problems occur when extreme views are expressed in the media without adequate background material, causing alarm in susceptible people who, for the most part, do not have a specific knowledge of the subject and cannot distinguish a legitimate lone voice from the general body of opinion on the subject. It is an unfortunate fact of life that people are very easily frightened by the media and often do not accept reassurance from those more familiar with the subject.

An even more worrying fact is that many people understand radio radiation as being synonymous with nuclear radiation, an aspect mentioned in respect of the public in Chapter 1. Since the latter is well known as a dangerous type of radiation, it can easily be seen why people continue to attribute the most serious effects of ionising radiation to radio frequency sources. The use of anonymous questionnaires at safety lectures has shown that such fears apply also to some technical people, including science graduates.

As a result, much time has to be given by RF safety specialists to explaining to technical people the difference between these two types of electromagnetic radiation and to explaining current views on RF radiation. It has to be said that reassuring people is not easy nor can it be totally authoritative since surprisingly little is known with any real certainty on the subject of RF radiation. Usually, the most that can be said is that there is no evidence to support a particular viewpoint.

It is interesting to note that other forms of electromagnetic radiation such as visible light and infrared radiation do not excite a great deal of public interest although photons of these have a much greater energy than RF radiation. The longest visible light wavelength has a photon energy of about 1.7 eV whereas the photon energy of the top end of the RF spectrum (300 GHz) is more than one thousand times less. Even ultraviolet radiation, for which the shortest wavelength portion of the spectrum at 100 nm and below is ionising radiation capable of quite serious effects, only invokes a modest amount of public concern.

The task of determining safe limits for RF radiation guides and standards is particularly difficult due to the lack of any substantive knowledge of the effects of RF radiation other than those concerned with direct thermal effects, shocks, burns and induced body currents at the lower frequencies. Most safety limits are determined by balancing benefits against risks and consequences and the latter are not well enough understood to permit this approach.

Notwithstanding such difficulties, standards for RF radiation safety are needed and practical safety limits for everyday work have to be set by some sort of consensus amongst those experienced in the field, having regard to such research as is available. If the limits are just set arbitrarily low, the use of RF power may become a serious practical problem, without offering any assurance that the low limits actually achieve anything.

It is not the purpose of this chapter to seek to resolve the differing views of the various factions interested in the debate on electromagnetic fields but rather to provide a broad outline of those known hazards of RF fields which are accepted by bodies concerned with producing safety standards, and to mention some of the areas of current investigation. A useful report which provides more details of the medical aspects, reviewing the findings of many research papers is the UK NRPB report NRPB-R240 [3].

Generally speaking the chapter is concerned with RF radiation down to about 10 kHz and does not therefore address the question of power frequencies (50 and 60 Hz). There is a great interest in the safety aspects of the fields from power frequencies and reference should be made to published papers on the subject including WHO publication number 25 [12] which deals with this and a number of other subjects, in addition to radio frequencies.

We can define the potential hazards of RF radiation in terms of:

1 Direct effects on people:
   (a) thermal effects attributable to the heating of the human body due to the absorption of RF energy. At lower frequencies this includes heating due to excessive current densities in some parts of the body.
   (b) shocks and burns which may result from contact with ungrounded conductors located in electromagnetic fields.
   (c) the so called 'athermal' effects where it is postulated that the fields act directly on biological tissue without any significant heating being involved.
2 Indirect effects on people wearing implantable devices such as heart pacemakers, insulin pumps, passive metal plates and other related hardware.
3 Effects on flammable vapours and electro-explosive devices, e.g. detonators (dealt with in Chapter 5).

Category 3 above may, of course, also involve people who may be present near the subject and may be affected by fire or explosion.

Some aspects of these topics may be differentiated in a general way in relation to the frequencies involved. The basic philosophy postulated in the IEEE C95.1–1991 standard [4], is that quasi-static considerations should apply at the lower end of the frequency spectrum and quasi-optical considerations at the upper end (above 6 GHz). The key factors are:

1 Below 100 kHz current densities induced in the human body and the electro-stimulation of tissues are considered to be the limiting factors.
2 From 0.1 MHz to 6 GHz, specific absorption rate (SAR) is the relevant factor.
3 Above 6 GHz, power density limits are used to control exposure.

In order to limit burns and shock at frequencies below 100 MHz, the permitted electric field strength is specifically limited. Below 300 MHz in this standard, both the electric and magnetic fields must be separately measured. Between 30 and 300 MHz there is a possible easement of the requirement if it can be shown by analysis that measurement of one of the fields is sufficient to secure compliance with the standard.

Note that other standards differ with regard to the frequencies at which electric and magnetic fields have to be measured separately. This is usually indicated by the absence of power density limits for those frequencies in the standard concerned, or their presence for information only.

Occupational and public safety limits

There is one general issue amongst those creating standards which results in strong differences in views. This is the question of whether separate limits are needed for these two groups. The wider issue of standards is dealt with in Chapter 4 but it is useful here to look at the arguments on both sides because they touch on harmful effects.

Some people feel that since there is no accepted concept of 'dose' for RF radiation such as exists for ionising radiation, there is no scientific case for separate limits for the two groups. (As a basic concept, dose = dose rate multiplied by the exposure time.) Consequently such people see the issue as a social and political matter. On the other hand some people believe that the duration of exposure is a significant factor in determining risks. It is true that, in general, populations feel protected if they are subject to tighter limits than those whose occupation requires them to be exposed. This is probably a universal feeling to which most of us would subscribe, especially if it relates to some occupation other than our own.

There could be a case on these grounds alone for lower limits for the public, though there are economic costs for such a decision. A factor often overlooked is the general acceptance by most bodies that the 'public' includes those non-technical personnel working for organisations using RF radiation. Thus there is a mixing of groups in employment and some sort of segregation is implicit.

The medical aspects raised include:

Members of the public include the chronic sick, including people with impaired functions such as the thermo-regulatory functions and who may therefore be subject to risks which might not apply to fit people.

The suspicion that RF radiation may have undesirable effects on people taking some types of drugs for medical conditions.

The fact that the athermal effects of RF radiation may eventually prove to have adverse effects on human health.

The possibility that RF radiation effects are cumulative, i.e., related in some way to 'dose'.

As it can be seen, these statements are of the precautionary type, the argument being that those who have to work with RF radiation choose to do so but the public in general have not made any such choice.

The fact that the arguments either way are not proven does not preclude the taking of a decision which is believed to err on the safe side, though the economic consequence is the cost involved in segregating the two groups, especially where the limitations of land ownership or occupation affect radiation levels at the interfaces with the public.

The IEEE C95.1–1991 standard, referred to earlier, tackles this problem by defining the need for RF radiation safety measures in terms of 'areas' rather than groups of people, namely 'controlled areas' and 'uncontrolled areas'. The former is an area where people who are knowledgeable about RF radiation are employed. The latter covers all the other employees. Extra safety factors are included for the last category. This concept of control by segregated areas broadly follows the practice for ionising radiation, though it would be very undesirable for the comparison to cause any confusion between the two types of radiation.

As can be seen from the discussion in Chapter 4, some standards only have one set of limits for all people, whilst others have separate provisions for occupational work and for the 'public'.

Specific absorption rate (SAR)

This term was used earlier and needs some explanation in the context of safety assessments. It is used to quantify the absorption of energy in tissue and is expressed in watts per unit mass of tissue, usually Wkg^{-1}. It is convenient to use the concept of the 'standard man' to aid discussion of the thermal aspects of RF radiation. The generally adopted standard man has a height of 1.75 m (5 ft 9 in), a weight of 70 kg (154 lb) and a surface area total of 1.85 m^2 (20 sq ft).

It is easy to see that the weight of the standard man is part of the definition of SAR so, for example, if it is known that the total power deposited in the standard man is 7 W, then the average whole-body SAR is 7/70 Wkg^{-1} or 0.1 Wkg^{-1}.

A 'worst-case' expression to relate specific energy absorption and temperature providing that the effect of cooling is neglected is given by the NRPB report [3] as:

$$T = J/(c \times 4180)$$

Where:
T = temperature rise (°C)
J = specific energy absorption (Jkg^{-1})
c = relative heat capacity (= 0.85)
Note also that J (Jkg^{-1}) = SAR (Wkg^{-1}) × exposure (seconds)

Hence a SAR of 2 Wkg^{-1} for 30 minutes will give a temperature rise of 1°C, neglecting cooling.

At very low frequencies (tens of kilohertz) energy absorption is relatively low. Absorption increases to a maximum at human resonance, which for adults is somewhere between 30 and 80 MHz depending on height and whether the person is effectively earthy or not. Above resonance absorption declines somewhat.

There is no practical way of measuring the SAR of a human being. In order to make calculations of SAR, either computer modelling or practical experiments with dummy persons using substances which simulate the electric characteristics of human tissues are undertaken.

Practical studies which simulate the human body use either standard shapes of hollow plastic objects such as spheres or hollow plastic human models generally known as phantoms. Their construction will depend on the temperature measurement technique to be used.

The most common systems are infrared (IR) scanning and temperature recording systems or the use of implantable temperature probes connected to some form of controller and data logger.

Phantoms in which implantable probes are used may be filled with a liquid or semi-liquid media simulating human tissue. This may be a homogeneous filling or elaborate layering and scaling may be done to represent the bones and organs of the human body with their different tissue simulations. The latter is obviously more expensive in time and materials, but can provide some differentiation of tissues.

In the case of phantoms to be used with IR thermography, the phantom can be bisected in the planes of interest, vertical and horizontal and flanges fitted to facilitate dismantling and assembly at the sections.

Again the simulation of tissue may be homogeneous or structured. The open faces of the sections are often covered with a close woven material which will ensure electrical contact of the two halves when assembled.

The complete phantom is exposed to a known uniform RF field for a specified time. The phantom is then split at the relevant sections and their open faces subjected to IR thermography to provide a plot of temperatures. In fact there will usually be two scans, one before the phantom is exposed and one after so that the temperature changes can be recorded.

Whichever type of system and phantom is used, the object is to calculate either whole body SAR or, sometimes, a local SAR. With probe systems, it is important that the probes should not perturb (distort) the RF field. A paper by Stuchly *et al.* [5] illustrates the scanning probe arrangement, using a non-perturbing probe system. Phantoms do not simulate the thermo-regulatory system of the human body and the results cannot be regarded as indicative of the temperatures likely in a live healthy human body.

Computer modelling attempts to model the human body by sub-dividing it into cells and attributing the relevant characteristics to each cell by analogy with the structure of a human being. There are limitations resulting from the deficiencies of any given model relative to a human body both in respect of the static model and the modelling of the dynamic performance of the complex thermo-regulatory mechanism of the human body.

The validation of computer modelling is difficult since it is generally only possible to compare it with some experimental trial such as the phantom method described above, despite the limitations of the method. Another paper by Speigel *et al.* [6] illustrates both a computer simulation and the comparison of the results with a phantom model.

Although one can identify the problems these methods pose, it has to be recognised that it has not yet proved possible to devise any other measurement method.

Thermal effects

General

There is general agreement that the main demonstrable effect on the human body is the thermal effect, i.e., the transfer of electromagnetic field energy to the body. A very high percentage of the human body is made up of water and water molecules are polar molecules liable to be influenced by impinging electromagnetic fields. Hence those tissues having a significant water content are most liable to be influenced by fields. Some other tissues also have large polar molecules

The effect of RF on such body tissues is to cause polar molecules to attempt to follow the reversals of the cycles of RF energy. Due to the frequency and the inability of the polar molecules to follow these alternations, the vibrations lag on them, resulting in a gain of energy from the field in the form of heat which causes an increase in the temperature of the tissue concerned.

With the widespread use of microwave ovens, most people have a practical awareness of the fact that microwaves can heat tissue, as represented by the animal tissues used in cooking, and should not find it too difficult to understand the nature of the thermal hazard. The amount of heating depends on the amount of energy absorbed and the activity of the human thermo-regulatory system. In turn, the amount of energy available depends on the power of the source and the duration of the exposure, 'cooking time' in the oven context.

Human thermo-regulation

In the healthy human body, the thermo-regulatory system will cope with the absorbed heat until it reaches the point at which it cannot maintain the body temperature satisfactorily. Beyond this point, the body may become stressed.

Excessive exposure can give rise to hyperthermia, sometimes referred to as heat exhaustion, an acute, treatable condition which, if neglected could have serious results. Excessive heating can also cause irreversible damage to human tissue if the cell temperature reaches about 43°C.

The author has never come across a case of hyperthermia in connection with RF radiation even with the highest power transmitters. This is possibly attributable to the commonsense of those who work with them. Nevertheless, with some equipment installations there is the potential for excessive exposure which, in the worst scenario, might have very serious consequences, so there is no room for complacency.

A rise in body core temperature of about 2.2°C is often taken as the limit of endurance for clinical trials [7]. For RF radiation purposes, a limit of a 1°C in rectal temperature has often been postulated as a basis for determining a specific absorption rate (SAR) limit for human exposure. Most western occupational standards are based on an SAR of $4\,\mathrm{Wkg^{-1}}$ divided by ten to give a further safety margin. Thus the general basis is $0.4\,\mathrm{Wkg^{-1}}$.

It should be noted that people with an impaired thermo-regulatory system or with other medical conditions which affect heat regulation may not be so tolerant to the heating permitted by standards which have been set for healthy people. Those taking some forms of medication may also be affected adversely.

There are also factors other than general health which affect the ability of the human body to handle heat energy. For example, a period of strenuous physical work can elevate the rectal temperature.

Another factor is the environmental condition – ambient temperature and relative humidity can make a considerable difference in the ability of the human body to get rid of excess heat.

Consequently, a given SAR may, for a constant ambient temperature and specified exposure time, give different body temperatures if the relative humidity is changed from a high figure, say 80%, to a low one, say 20%. Put the other way round, a specific increase of rectal temperature of, say, 1°C will require a much higher SAR at low relative humidity than is needed at high humidity.

In 1969, Mumford[8] identified this aspect and proposed a 'comfort index' whereby the higher safety level then in use ($100 \, \text{Wm}^{-2}$ for all the frequencies covered) was reduced as his temperature-humidity index increased. Current standards such as the IEEE C95.1–1991 standard claim to accommodate environmental factors in the large contingency allowance put into the permitted limits.

A particularly interesting paper on the thermo-regulatory mechanisms of the human body is that of Adair[9]. The paper describes the regulatory mechanism in some detail. It notes experimental work done to establish the thermal equivalence of heat generated in the body during physical exercise and passive body heating such as that from HF physiotherapy equipment.

It also makes reference to the radical difference between the thermal responses of man and various animals and the consequent difficulty in extrapolating animal exposure data to human beings on this account, quite apart from any resonance differences.

RF penetration in human tissues

In considering the amount of energy absorbed by the human body, it is necessary to recognise that the percentage of incident radiation which is actually absorbed depends on frequency and the orientation of the subject relative to the field.

In human tissues, RF radiation may be absorbed, reflected or may pass through the tissue. What actually happens will depend on the body structure and the tissue interfaces involved. These interfaces are the transitions from tissue to tissue or tissue-air-tissue and are clearly complex in the human body.

The depth of RF penetration of the human body is also an important factor. In the HF band, the deeper penetration is used for diathermy treatment where the deposition of heat is intended to have a beneficial effect on that part of the body considered to need treatment. The deep deposition of RF energy needs to be carefully controlled to avoid damage to tissues which might not be noticed by the subject due to lack of sensory perception of heat in the organs concerned.

The measurement of the RF characteristics of human tissue can, for the most part, only be done with chemical simulation of tissue, since there are problems with the use of excised human tissue for this purpose. The penetration depth is usually given as the depth where the incident power density has been reduced by a factor of e^{-2}, i.e., down to about 13.5% of the incident power density.

The penetration decreases as frequency increases. Figure 3.1 has been drawn using some of the data from published tables, the work of Schwan,

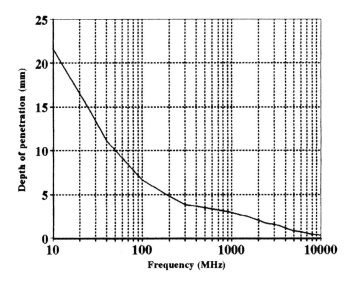

Figure 3.1 *Depth of penetration of RF energy in tissues with high water content (data from reference 10).*

Cook and Cole and other researchers. The tables are given in a paper by Johnson and Guy [10]. It illustrates laboratory calculated penetration depths versus frequency for tissues with high water content.

The illustration should only be considered as giving a rough picture of the change of penetration depth with frequency as the laboratory determination of these data is subject to various factors including temperature dependency. Tissues with a low water content have significantly deeper penetration.

At the microwave end of the RF spectrum, deposition of energy is confined to the surface layers of the skin. The penetration depth at the higher microwave frequencies may only be a few millimetres[11]. Deposition of energy in the surface layers of the skin may lead to thermal injury, the risk increasing as the frequency increases.

Resonance

It has been seen how the weight of the standard man is linked to the use of the concept of specific absorption rate. The purpose of a standard height may be less obvious. To see the effect of this it is necessary to consider how the absorption of energy is affected by frequency.

It is also necessary to define the attitude of the model relative to the plane wave field to which it is subjected.

Figure 3.2 shows the average SAR in a spheroidal model man subjected to a field of 10 Wm^{-2} and displayed in three curves[12]. The curves are labelled

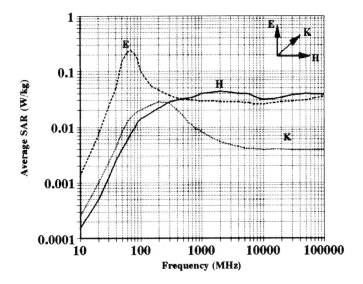

Figure 3.2   *SAR versus frequency for orientations parallel to the E, H, and K vectors*
(Courtesy the World Health Organisation, European Office)

E, H and K and indicate that the model man was successively orientated parallel to the electric field (E), the magnetic field (H) and the direction of propagation, head to toe (K).

Considering the curve E, it can be seen that absorption is lowest and declining rapidly with decreasing frequency at 10 MHz. It increases rapidly with increasing frequency to peak at about 70 MHz. The peak represents 'resonance' of the model man.

Put simply, this means that at the resonant frequency, the absorption of RF energy is at a maximum. The reader unfamiliar with electrical resonance may be familiar with mechanical resonance where at a given frequency some object which is excited with constant power over a range of audio frequencies, manifests a large amplitude of vibration at one particular frequency. The energy involved may then result in acoustic noise and possibly the eventual fatigue of materials.

Returning to the RF resonance case, the resonant frequency is related to the height of the erect person. Resonance occurs when that height corresponds to approximately 0.36 to 0.4 wavelengths.

Hence using 0.4 for the standard man, $\lambda = 1.75/0.4 = 4.37\,\text{m}$ and the frequency is approximately $300/4.37\,\text{MHz} = 66\,\text{MHz}$.

If the subject is effectively earthy due to bare feet or conductive shoe material, the resonance will occur at half the above frequency, i.e., about 33 MHz. Table 3.1 gives a few examples for subjects who are non-earthy,

Table 3.1   *Effective resonant frequency for an erect person in a vertically polarised field versus height using the relationship h = 0.4 λ*

Subject height (metres)	Resonance (MHz)
0.5	240
0.75	160
1	120
1.25	96
1.5	80
1.75	68.6

using the 0.4λ calculation. Small children obviously resonate at higher frequencies and tall adults at lower frequencies.

Hence a frequency band can be established which covers all people, large and small. Strictly, the occupational frequency bandwidth for resonance is smaller since the range of heights for employed people is smaller.

Gandhi[13] states that at resonance a human being absorbs energy 4.2 times greater than that which might be expected from consideration of the physical cross-section of the body. Further, when the person is effectively earthed, the resonant frequency is reduced to approximately half of that for the non-earthy condition and the energy which is absorbed is about 8 times that expected from consideration of the physical cross-section. Another way of conceiving this is that the effective electrical cross-sectional area of the exposed person is several times that of the actual cross-sectional area at resonance.

It will be noticed in Figure 3.2 that the electric field curve (E) indicates that the electric field gives rise to more absorption than the magnetic field curve (H) up to about 700 MHz, the difference being considerable over much of that frequency range.

Hot spots

The human body is made up of a mixture of types of tissue, for example, skin, blood, bone, muscle and fat. When the human body is exposed to RF radiation, there is, as described earlier, some degree of absorption of the energy in the form of heat. However, the absorption of RF energy in the human body which is made up of such a complex mixture of tissues, can result in a non-uniform distribution of heat. Hot spots (high local SARs) may occur in the human body over the range of about 30 to 400 MHz. These hot spots will be evident at frequencies around body resonance where absorption is greatest and at sub-resonances in parts of the body. Gandhi[14] gives the adult human head resonance range as being of the order of 350 to 400 MHz with a volume–averaged SAR of 3.3 times the whole-body SAR at resonance and the absorption cross-section as about three times the physical cross-section.

He also gives some local SAR values for knees, ankles and the neck for body resonance in the grounded man (about 34 MHz) and the ungrounded man (about 68 MHz).

The measurements were made with scaled human phantoms and showed hot spots at the knees, ankles, elbows and, in the case of the non-earthy model, the neck. These have some 5 to 10 times the average whole-body SAR.

It is difficult to tackle the problem of non-uniform heat absorption by seeking to identify the location and temperature of such hot spots. Using physical models poses the problem of carrying out measurements without affecting the distribution and magnitude of the effects due to the presence of the measuring devices and, as mentioned previously, physical models cannot simulate the human thermo-regulatory system.

Work has been done on the subject of 'hot spots' using computer modelling but this again poses the problem of validating such models as being an adequate and correct representation of the functioning human body. The reason that attention has been given to this problem is a simple one. If a safety standard defines a safe power density limit for a particular frequency on the basis of the average whole-body SAR but some small parts of that body reach significantly higher temperatures than others, there must be concern as to whether these can be harmed in some way.

Some high ratios between mean body temperatures and hot spot temperatures have been noted[3]. Ratios suggested from experiments using magnetic imaging range from 10 to 70, though this reduces to a factor of 2 to 4 when the SAR is averaged over individual organs.

The theoretical end point could be where the hot spot is so hot as to cause cell damage, in which case it would be necessary to adjust the average permitted levels to reduce the hot spot temperatures. It has to be said that little is known about the real effects in a healthy individual with an efficient thermo-regulatory system as contrasted with computer or model simulation.

A paper by Gandhi and Riazi[11] looks at the power capabilities of RF sources in the frequency band 30 GHz to 300 GHz and identifies the possibility of high energy deposition rates for the skin at frequencies in that range due to the very shallow penetration depths. It also looks at the possibility that dry clothing may act as an impedance transformer, increasing the amount of energy coupled into the body. The thickness of clothing in this frequency band is a significant fraction of the incident wavelength. This could, for a given incident power density, exacerbate the situation by further increasing the deposition in the superficial layers of the skin.

The IEEE C95.1-1991 standard recognises the problem of energy deposition in the superficial skin areas by progressively reducing the averaging time for exposures at above 15 GHz from the usual 6 minutes to a shorter period.

Susceptible organs

From the thermal transfer point of view, the two organs which are considered more susceptible to heat effects than others are the eyes and the male testes. Neither of these have a direct blood supply and hence do not have that means of dissipating the heat load.

Effects on the eyes

The production of cataracts in animal experiments using RF has been well established. It is generally considered that this effect is a thermal one. Experimental work has been limited to animals and the different physical characteristics of the eye structure in different types of animal do give rise to different results. Also, the depth of penetration of the eye tissues is dependent on the frequency of the radiation.

It is thought that for human beings the frequencies most likely to cause cataracts lie between 1 and 10 GHz and probably require power densities of $1000\,Wm^{-2}$ to $1500\,Wm^{-2}$. Whilst it is easy to do animal experiments with small localised fields, in practice, people exposed to RF fields related to antenna systems are likely to experience whole-body radiation and these sort of levels for whole-body radiation are far in excess of those permitted for microwave work.

Some recently reported work claims that microwave radiation at low levels, particularly with pulsed radiation, can affect susceptible parts of the eye. Gandhi and Riazi[11] referred to experiments on rabbits at 35 GHz and 107 GHz where some eye damage (albeit reversible) had been sustained with a total absorption in the eye of 15 to 50 mW.

They suggested that at millimetric wavelengths, the power absorption of the human eye might be of the order of 15 to 25 mW for an incident power density of $100\,Wm^{-2}$ after 30 to 60 minute exposures.

It was considered that studies of exposures longer that 60 minutes are needed to investigate this. If the estimates are found to be true, there will be some reason for concern about the new power density limits for these frequencies in the latest standards.

From other sources, some low thresholds for harmful effects have been reported for cases where a substance used for eye treatment (timolol maleate), had been applied. If this work is subsequently confirmed by others, the whole subject may need further investigation.

In the author's view there is often unnecessary exposure to the eyes, for example by holding the head close to open RF amplifier circuitry when aligning or diagnosing on a bench without bothering to switch off. These can usually be avoided by better safety disciplines and by the use of modern optical aids including optical fibre inspection equipment.

There is similarly a need for caution in working with RF radiation so as to avoid the unnecessary eye exposure which can sometimes occur where waveguide flanges are removed without the source being switched off, and worse still, by the silly practice of looking down such waveguides. Although people think that the old practice of looking down the waveguide with power on to look at the electronic tube stopped long ago, cases have arisen as recently as three years ago!

Effects on the testes

Experiments with anaesthetised mice and rats showed[3] that male germ cells are depleted by exposure to SARs of about $30\,Wkg^{-1}$ and 8–$10\,Wkg^{-1}$, respectively. Conscious mice exposed to $20\,Wkg^{-1}$ and $9\,Wkg^{-1}$ respectively, did not show any effect. The difference is regarded as being due to the fact that the anaesthetised animals were not able to regulate their testicular temperature. Other studies with rats reported a transient decrease in fertility with an SAR of about $6\,Wkg^{-1}$.

There seems to be little if any published information regarding such problems with human adult males. It seems likely that the whole-body SAR required to produce a sufficient temperature increase in the testes of an adult male would produce some basic signs of warmth and discomfort, resulting in withdrawal of the subject from the RF field. This is, of course, purely speculative since the author is not aware of any research carried out with men. However, with many years working in a large organisation manufacturing high power transmitters, no complaint of this kind has arisen.

Hearing effects

It has long been known that some people can 'hear' the pulse repetition frequency of radars and similar equipment. In this field it is not usually difficult to find human volunteers for tests so that there is no problem of relating other animals to people.

It is therefore surprising that more work has not been done in this field. The work of Frey[15] reported in 1961 involved tests with volunteers using two transmitters of frequencies 1.3 GHz and 2.9 GHz, the former being pulsed with a 6 μs pulse (244 Hz repetition rate) and the latter with a 1 μs pulse (400 Hz repetition rate).

He gave the mean power density threshold of hearing for those able to hear anything as 4 Wm^{-2} and 20 Wm^{-2} respectively. The corresponding peak pulse power densities were 2.6 kWm^{-2} and 50 kWm^{-2}. It was stated that the human auditory system responds to frequencies at least as low as 200 MHz and at least as high as 3 GHz.

In another paper, [16] Frey reported that the sounds heard included buzzing, hissing, and clicking and depended, among other things, on the modulation characteristics. In these tests Frey used a frequency of 1.245 GHz. A constant repetition rate of 50 Hz was used and longer pulse widths (from 10 to 70 μs).

The pulse width was changed to adjust mean power and peak power densities. The volunteers were required subjectively to assess the loudness of the sound heard relative to a reference sound which had been transmitted.

The general finding was that the perceived loudness was a function of the pulse peak power density, rather than the average power. The peak power density for perception was less than 800-Wm^{-2}.

The nature of the effect has been the subject of much investigation[17]. It seems generally agreed that the pulsed RF energy causes an expansion in the brain tissue due to the small but rapid temperature change involved.

This causes a pressure wave which is transmitted through the skull to the cochlea where the receptors respond as for acoustic sound. It is not necessary to have the middle ear intact. The temperature increase which causes the pressure wave is considered to be less than 10^{-5}°C.

It is perhaps worth noting that sometimes the pulse repetition frequency of high power radars can be heard from objects such as old wire fencing, and this can easily be confused with the above phenomena. Presumably the effects on old fences involves some form of rectification of the RF currents due to corroded junctions within the fence, and the consequent vibration of some fence elements at the pulse repetition rate.

Although the results of laboratory tests have been published, little, if anything, seems to have been published in recent times regarding the practical experience of those working on transmitting sites and any problems they may have noticed. With the low levels mentioned as thresholds for hearing, it might be thought that many radar personnel would experience this phenomenon.

A survey was carried out by the author across 63 engineers working with the transmitting side of radar. Many of the participants had 30 years or more experience in that work. The survey has no scientific basis, being limited to the collection of anecdotal evidence from those concerned by means of a questionnaire.

The results were interesting in that only three people claimed to have heard the pulse repetition frequency (or sounds related to it) and for two of these, each cited only a single experience in unusual circumstances. Both occurred on a customer's premises, during the Second World War.

Both of these people considered that the circumstances led to exposure to very high fields but there was no measuring equipment available in those days and in consequence, little safety monitoring. The third case was interesting in that it seemed to imply a different mechanism, one which has occasionally been reported in the past. This person claimed that he had heard the pulse repetition frequency on a customer's premises and attributed this to a tooth filling. (The Frey work in 1961 did include the use of shielding to exclude the 'tooth filling' possibility.)

It was further claimed that this ceased when the tooth was extracted. Strangely enough, another person who gave a negative answer to the basic question did claim to hear a local radio amateur when at home, again attributing this to a tooth filling with the same claim that it ceased after the tooth was extracted!

Outside of this survey, there was one engineer in the same organisation who regularly claimed to hear the pulse repetition frequency on company premises, but was able to live with it. This applied in an environment which was maintained within ANSI C95.1-1982.

This account is again anecdotal, the experience extending over a number of years. It was the only case in which this phenomena occurred on company premises.

Assuming that none of the respondents to the questionnaire had chosen to suffer in silence, this particular company, which designs and manufactures high power civil and military radars, does not seem to have a problem with auditory effects despite the high radar peak powers usually involved and the fact that much of their high power work lies within the 0.8 to 4 GHz frequency range.

Limb currents

Up to about 100 MHz, theoretical consideration of currents induced in the human body and especially the limbs, has given rise to some concern. As a result, research has been carried out to ascertain the magnitude of RF currents induced in the human body. It has been established that currents in the legs of an adult in an RF field may give rise to large SAR values at places where the effective conductive cross-sectional areas are small. Hence, the current density will be much larger than that implied by consideration of the actual cross-sectional area at that place. The knee and the ankle are examples of such areas, and some attention has therefore been given to SAR values associated with them, particularly the ankle.

Gandhi and Chen[18] have shown by measurements made with people that the induced currents are highest when the human body is erect and barefoot, i.e., earthy, and parallel to a vertically polarised plane wave field. The leg currents are proportional to frequency and to the square of the height of the person exposed.

An approximate formula (up to about 27 MHz) for the current in the leg of an erect barefoot person where the electric field is vertical is given as:

$$I \text{ (mA)}/E \text{ (Vm}^{-1}) = 0.108 \times h^2 \times f \text{ (MHz)}$$

where h = subject height (m), f = frequency (MHz) and E = electric field (Vm^{-1})

Example:

For a field E = 60 Vm^{-1}; f = 1 MHz; h = 1.75 m:

$$I \text{ (mA)} = (0.108 \times (1.75)^2 \times 1) \times 60 = 19.85 \text{ mA}.$$

Above 27 MHz, the measured currents peaked around 40 MHz, reflecting resonance of the subject. An empirical expression for current above 27 MHz which includes an element representing the resonance frequency of the 'standard man', gives a sine wave shape for I/E versus frequency.

At field limits corresponding to the ANSI C95.1-1982 standard electric field limits[19] for frequencies from 3 to 40 MHz, values of SAR at the ankles of 182 to 243 Wkg^{-1} have been reported from tests carried out. Ankle currents were reduced when footware was used, depending both on the frequency and on the electrical properties of the material of which the footware was made. The current when wearing shoes ranged from 0.8 to 0.82 times the barefoot current. Previous work at 1 MHz showed corresponding fractions as 0.62 to 0.64. The increase with frequency is due to the fall in the impedance to ground as the frequency increases.

Another paper by Chen and Gandhi[20] illustrates the task of computer modelling the human body in order to establish the RF currents induced over the frequency range 20 to 100 MHz. The results of the calculations are presented in a series of graphs. The results are said to agree with experimental data produced by other workers.

RF safety standards now include limits for the values of leg currents up to 100 MHz as well as contact current limits to prevent shocks and burns.

RF shocks and burns

At low frequencies and up to about 100 MHz, contact with passive objects in RF fields may result in currents flowing through that part of the body in contact, usually the hands, causing shock and sometimes burns.

These effects can result from contact with almost any conductive object such as fences, scrap metal, unused dish and similar antennas or other equipments stored in the open, vehicles, farm machinery, metal buildings, etc.

Burns may result when the current density ($mAcm^{-2}$) is excessive due to the contact area being relatively small. The possibility of a burn is reduced with the greater area of a full hand grasp. However, this is rather academic since contact is usually inadvertent and often involves the finger tips.

A paper by Chatterjee *et al.* [22] deals with the measurement of the body impedances of several hundred adult subjects, male and female, over the frequency range 10 kHz and 3 MHz. Experiments were also carried out on threshold currents for perception and for pain. It was generally found that up to 100 kHz the sensation experienced was that of tingling or pricking and above 100 kHz the sensation was one of warmth. The calculated current for contact with the door handle of a van of effective area of $58 m^2$ and 0.5 m effective height at 3 MHz is given as 879 mA when the field is $632 Vm^{-1}$.

Because of the fact that burns essentially result from the current density at the point of contact and hence the effective contact area, it is quite possible to experience currents exceeding a given standard without incurring burns, purely as a result of a fortuitously large contact area. It is clearly important to measure contact currents rather than operate on the practice of assuming that if no burn occurs then there is no hazard.

It should not be overlooked that parts of the body other than the hands may incur burns. A common example is where shorts are being worn, as the bare leg may contact metal objects. People have also been known to sit on metal objects when wearing shorts.

Present views on a limit for occupational exposure to contact current over the frequency range 0.1 to 100 MHz seem to range from 100 mA to 20 mA. Touch burns have been predicted as possible at 60 mA with a contact area of $0.2 cm^2$.

Apart from the undesirability of incurring shocks or burns, there are other indirect effects which, in the writer's experience, can be more worrying. Quite small shocks incurred by people working on structures can result in an involuntary movement (startle response) and a possible fall. Current safety standards are not likely to prevent these small shocks.

In a number of cases where a contract to fit new antenna systems on working sites has been involved, construction staff have started assembly and noticed 'sparks' between tools and the structure. Consequently, there has been considerable alarm because of the general fear of radiation and because of the awareness of the possibility of 'startle response' accidents.

In a number of these situations, the result has been a temporary cessation of work. The subject therefore needs more attention than the mere observance of the provisions of a safety standard regarding contact current since the probability is that these apparently minor manifestations may occur at lower limits and have indirect physical and psychological effects.

There will often be two cases to consider as far as contact with conductive material in an RF field is concerned. The first case involves the safety of people employed on a site or on company property and in most countries there are provisions for health and safety at work. The problem lends itself to good safety management which should include the prohibition of the dumping of metal objects in such fields. The second case is that of the public who own or have a right of access to adjacent land which may be irradiated by RF. Here those who own or lease such land have a right to use and deposit conductive objects where they wish and there is a duty on those responsible for the RF emitter to ensure that those people cannot receive shocks or burns from objects on their own property.

It may even be necessary to reduce the radiated power or change the antenna characteristics in order to ensure that this is so. The same applies to land on which the public may have a right of access, possibly with bicycles and cars. Vehicles in particular can give rise to significant contact current when in an RF field and since there are often a lot of them about, this can be a significant problem even if it only causes annoyance rather than real injuries.

Perception of a sensation of heat in RF fields

As mentioned earlier in this chapter, RF energy in the higher frequencies of the RF spectrum above perhaps 3 GHz can be detected by temperature sensors in the skin since, as the frequency increases, the energy is increasingly deposited in the outer layers of the skin. The indications resulting are dependent on a number of factors including differing heat sensitivity in different parts of the body, the duration of the exposure and the area exposed. WHO[12] notes that in tests on mammals the threshold temperature for cellular injury over seconds to tens of seconds (42°C) was found to be below the pain threshold (45°C). For this reason, the avoidance of the skin heating sensation does not provide a reliable protection against harmful exposures.

It is postulated that it may be a better indicator at frequencies of tens of GHz and higher where the wavelength is comparable with or less than the thickness of the skin. Prudence seems to suggest that it should be completely disregarded as an indicator until much more is known about all the variables involved.

There is a universally accepted statement about RF radiation work that people should not remain in a field which gives rise to a sensation of warmth even if the power density is within the permitted limits of a standard. It will be evident from the previous discussion that this does not guarantee that no harm has been incurred but is intended as an extra warning.

At frequencies well below those being discussed, the penetration of RF in the human body (page 49) is such that much more of it will be below the skin sensors and there may not be any physical sensation of warmth. The prevention of internal damage has to be by the limitation of exposure.

Pulsed radiation

Pulsed RF radiation is very common, typically in radar and in some data transmission systems. Pulse transmission is discussed elsewhere but the two aspects relevant here are the mean power density and the peak pulse power density. A typical radar may have a factor of 1000 between these so that, say, a mean power density of $50\,\mathrm{Wm}^{-2}$ would imply a peak pulse power density of $50\,\mathrm{kWm}^{-2}$.

In the early days of RF radiation safety regulation, emphasis was on mean power and this left open the question of limiting the peak power density. As a result of concern expressed by some people, research has been undertaken to examine possible effects attributable to high peak power pulses.

Experiments with animals have indicated that the startle response to a loud noise was suppressed by a short pulse of microwave radiation. Body movements have also been induced in mice. Calculations suggest a large SAR value in these animals.

It is considered that pulsed radiation may have specific effects on the nervous system. Most standards now place a limit on peak pulse power density and pulse energy and this is discussed in Chapter 4.

Athermal effects of RF radiation

This term is used to describe any effect which is thought to arise by mechanisms other than that involving the production of heat in the body. It has been somewhat controversial, some people disputing whether such effects existed. However, most people now probably accept the need at least to investigate observations which do not seem to be linked to the thermal deposition of energy in the human body.

With regard to tumours, there seems to be a degree of consensus amongst most bodies.

This is probably best expressed in the UK NRPB press release on the report of the Advisory Group on Non-Ionising Radiation[23] which stated: 'We conclude from a review of all the evidence, including both that relating to humans in ordinary circumstances of life and that relating to animals and cells in the laboratory, that there is no good evidence that electromagnetic radiations with frequencies less than about 100 kHz are carcinogenic; this includes those produced by electrical appliances, television sets and video display units. With higher frequencies there is room for more doubt, some laboratory evidence suggesting that they may act as tumour promoters, although in this case the effect may be secondary to local tissue heating.' It goes on to recommend further research on the subject. A similar statement is made[12] in the World Health Organisation (WHO) publication No. 25.

Needless to say, there are people who do not subscribe to this view and since there are so many uncertainties reflected about the role of electromagnetic fields, if any, in cancer promotion, it is not possible to refute such views. The situation is often coloured by the fact that whilst we are all subject to the possibility of incurring this disease, some of those engaged in electrical and radio work who suffer the disease are inclined to attribute cancer to their occupation.

The topics encountered under the heading of 'athermal' effects cover almost everything to do with the human body. Reports and papers are very technical, requiring considerable practical familiarity with the subject matter. They range from the possibility of RF causing cancers as mentioned above, through the operation of all the systems and constituents of the human body, cells, tissues, organs, the immune system, reproduction, etc., to the psychological aspects claimed by Russian researchers.

These are not discussed further here but competent discussion of some or all of these topics can be found in references 3, 12 and 23 and in the standards and other national documents of various countries.

Effects on people wearing implantable devices

There are a number of implantable devices, active and passive, which are fitted into the human body. Perhaps the most common one is the heart pacemaker on which many people depend. There are two basic types of heart pacemaker. The first could be described as a demand pacemaker which will make up for missed heart beats as needed. The second type is the fixed pacemaker which operates continually at a fixed rate with no other form of control.

It is possible that some sources of RF radiation could interfere with the operation of pacemakers, the significance of such interference depending on the type of pacemaker fitted. The potentially more serious consequences of interference relate to interference with the fixed rate pacemaker. However, the two descriptions above are basic. With current developments in electronic devices there is always the possibility of the use of more sophisticated devices and the possibility of new problems of vulnerability to interference.

Many of these pacemakers are subjected to interference (EMC) testing by the manufacturer but the relevant information does not normally get communicated to the wearer of the device. Consequently, those responsible for the operation of RF transmitters and similar sources who may become involved with visitors wearing a heart pacemaker have no means of carrying out their responsibilities for the safety of such people.

The only recommendation that can be made is that such sites should have a sign requiring visitors to notify the manager that they are wearing a pacemaker. They can then be excluded from RF fields. A similar problem can occur at exhibitions where equipment is being demonstrated and where many people may be present.

There are other devices such as insulin pumps which are implanted and the views of medical authorities may need to be sought on these and any new types of implanted devices.

In the EEC there is a new Directive on Active Implantable Devices[66] but the current draft does not fully tackle the problem of the electrical characterisation of devices in terms of interference testing. There are also many types of passive devices fitted in the human body. These may include metal plates, rods and fixings. There is always the chance of these being resonant at the frequency in use at a particular site.

The author was involved in a case in the Middle East where a guard with a silver plate in his leg complained that it got hot when he was near a particular microwave transmitter. Such cases may not carry the same risks as that of pacemakers but can cause serious worries for people. For those employed with RF radiation, it seems desirable to record any such implants fitted when personnel are first employed and thereafter, should the situation arise. It is then possible to exercise supervision over the exposures to RF of such people.

In summary, the situation on all types of implantable devices is a dynamic one in which there is constant innovation. It may be necessary to ensure that surgeons and physicians have some understanding of the implications for those involved in RF radiation, so that their patients can be given meaningful advice.

Beneficial effects

Discussion of the effects of RF radiation on people would not be balanced without a brief reference to the beneficial effects which have been and are being applied in the medical field. Some aspects of this are discussed in Chapter 2 where a typical HF radiotherapy machine is illustrated.

Some recent uses are, in summary, as follows:

Bony injuries

There is considerable evidence that the application of RF energy at the site of a fracture speeds up the healing of both soft tissues and bony injuries. This is now fairly well established as a technique, though the mechanism by which such healing takes place has not been established with any certainty.

Treatment of malignant tumours

If cancer cells can be heated rapidly enough to the cell thermal death point, they can be destroyed. Microwave energy lends itself to application for this purpose. Current use is generally in association with other treatment, chemotherapy (cytotoxic drugs) or radiotherapy (ionising radiation). Getting the RF energy to the tumour site can be a problem and care is needed to avoid unnecessary damage to healthy cells.

Other organs

Techniques have been described which enable the application of RF to the male prostate gland to shrink the gland. Surgery is not needed and patients can usually return home the same day.

RF radiation effects summary

The induction of RF energy in the human body in the form of heat is accepted though there is a practical problem of establishing the amount of heat energy and its distribution in the body, due to the difficulties involved in direct measurement.

The susceptibilities of particular organs are known from considerations of the nature of the human body but practical experimental work is confined to animals with the consequent difficulty of extrapolating findings to human beings.

There is a body of knowledge on the subject of induced body currents since it is practicable to use volunteers for such work and measuring techniques have become available. There is also some knowledge of aural effects from human experiments.

Relatively little is known with any certainty about other possible effects, including those described as athermal effects, and much more research may be necessary to improve our knowledge in the face of the many difficulties, both technical and financial.

There is a body of knowledge, based on experimental work, for those indirect effects on people such as the inadvertent ignition of flammable vapours and electro-explosive devices. The information in respect of flammable vapours could easily be made into an international standard.

With the finite financial limitations on research across the world and the permutations of frequency, amplitude, modulation methods, etc., which might need to be explored, it seems necessary that such research should be organised and directed systematically to those topics which experts in the field feel to be priorities. In this way it might be possible to avoid the apparent randomness and fragmentation of present research and include provision for the independent replication of any seemingly important research results.

4

The development of standards for human safety

The nature of standards

In order to provide guidelines for RF safety it is necessary to try and define safety limits which will reflect those findings of researchers in the field of RF safety which have been accepted by governments or standards bodies. Those bodies which have had some recent influence include:

1 The American National Standards Institution (ANSI). The C95.1 committee was until recently operated under ANSI but now operates as a committee of the American Institution of Electrical and Electronic Engineers (IEEE).
2 The International Non-ionising Radiation Committee (INIRC) of the International Radiation Protection Association (IRPA). This body has operated for a number of years. In May 1991, IRPA announced the establishment of an International Commission on Non-ionising Radiation Protection, the function of which will be the investigation of the hazards that may result from all forms of non-ionising radiations. IRPA is an association of professional societies concerned with ionising and non-ionising radiation protection and states that it is non-governmental and non-political.
3 The UK National Radiological Protection Board (NRPB). The NRPB acts as Statutory Adviser to the Health and Safety Commission on both ionising and non-ionising radiation.
4 The Commission of the European Communities. The EEC Commission is currently drafting requirements for radiation protection involving the Directive for the Protection of Workers against the risks from exposure to Physical Agents and the Machine Safety Directive. Radiation here covers all forms of ionising and non-ionising radiation.

In addition, many countries have their own standards. The current situation on standards is that much change is taking place. Indeed, most of the organisations listed above are, or have recently been, involved in the revision of their standards. In particular, the current EEC drafting is likely to result in a standard which will apply in twelve or more European countries, thus providing some broad standardisation in Western Europe, in the absence of any recognised International standard.

Although a number of standards are described in this chapter, the main purpose of this is to illustrate the nature of standards and the aspects which are considered when they are drafted. The actual standards which apply to readers may well be changing now or, as in the case of the EEC countries, still being created.

RF radiation standards are needed for three purposes:

1 Control of both occupational and public exposure to RF electromagnetic fields.
2 The prevention of the ignition of flammable vapours and electro-explosive devices (EEDs) by RF energy.
3 The reduction of interference from sources of RF.

It should be noted that (2) above is not normally covered by RF radiation safety standards dealing with the exposure of people. Instead, most countries have their own separate standards or guides to cover the potential ignition hazards of flammable vapours and EEDs. In addition, the military forces usually have their own standards to cover the risks peculiar to the military field, especially military explosive devices. The general calculations involved for this topic are dealt with in Chapter 5.

Similarly, (3) above is covered by radio frequency interference (RFI) specifications or, to use the more modern and more comprehensive term, electro-magnetic compatibility (EMC) specifications. The latter takes into account both the production of RF interference and the susceptibility of equipment to interference.

There are established standards in this field based on the recommendations of the International Special Committee for Radio Interference (CISPR). This topic is not discussed in depth in this book but the problems of interference are mentioned where appropriate. A useful reference book on the subject which includes coverage of the European EMC Directive, is given at reference 21.

There is no useful relationship between the limits set in standards for the exposure of human beings and those which can cause radio interference since the latter must depend on the susceptibility of the equipment subject to the interference. Indeed a power density safety level of say $10 \, \text{Wm}^{-2}$, as set by a particular standard for a band of frequencies, corresponds to a plane wave electric field strength of $61 \, \text{Vm}^{-1}$, a level which is quite likely to be capable of interfering with sensitive circuits such as receiving and similar equipment.

In controlling human exposure to RF radiation there are two readily identifiable forms of safety standards reflecting whether the potential hazard relates to:

1 Adventitious (unwanted) radiation in the form of leakage from RF sources. Standards dealing only with this aspect are referred to generally as leakage or 'unwanted emission' standards.
2 Intended radiation from an antenna, machine or applicator, the standards then being referred to as exposure standards.

Leakage standards

Leakage standards generally set a maximum permitted radiation level at a defined distance from the surfaces of a source or product under specified conditions of use.

Hence the microwave oven has a specification for leakage at a specific distance, usually 50 mm but dependent on the standard concerned. This may be defined by standards or by legal provisions. The tests are easy to define and carry out and provide pass/fail quality control. In this sense they are easy to use as a product standard. For box-type products such as microwave ovens, RF test sources of significant power and other similar items, leakage measurements are needed. A similar approach for X-ray radiation leakage from television sets and visual display units (VJ⁻) is used in the ionising radiation field, the distance and maximum leakage dose rate being specified.

For transmitting equipment leakage in the broadcast and television fields, the IEC standard IEC 215 (1987)[24] and also BS 3192:1987[25] which is identical, specify a leakage limit, though this has not been updated in line with current standards. It does seem to be commonly the case that specific leakage limits in specifications generated some time ago can be overlooked and remain unrevised.

More commonly transmitters of all kinds are tested to in-house standards and, where applicable, to the purchaser's own specifications. Specifications from purchasers of RF generating equipment exhibit considerable variation in requirements and measurement methods. Where there is no specific leakage test prescribed for a product, the permitted levels of the relevant national or other RF exposure standard are generally used, with measurement mostly being undertaken at the standard distance of 50 mm.

For RF sources used on dedicated sites such as transmitting stations, equipment leakage will mostly affect the equipment user, since only those in close proximity to such equipment will be exposed to leaks which normally extend for relatively short distances.

On the other hand, the use of mobile or other equipment in the public domain is quite a different proposition since the public may be affected by leakage from the equipment, both in the safety context and also in respect of any interference problems.

In some cases the product concerned may be on domestic sale and thus used by the public generally, as is the case for some mobile radio equipment and for amateur radio transmitters.

In all cases where there is intentional radiation there is, by definition, the possibility that people, including the public, at a distance from the source, may be affected by the field of the wanted radiation.

Those employed with the source of radiation may, of course, be at greater risk owing to their closer proximity. Whilst leakage measurements take care of those who work in very close proximity to sources of RF, there is clearly a need for a standard for the exposure of any person from intentional radiation whether they are in the public domain or employed with RF radiation.

Exposure standards

It will be seen from the foregoing that there is no philosophical distinction between leakage and exposure standards. The limit values used may be identical and equipment which does not have its own leakage specification will, as previously noted, often be measured to the limits given in the relevant exposure standard. The practical difference is the arbitrary determination of the measurement distance used for leakage standards, that is to say, the distance between the sensor of the measuring instrument and the outer surface of the source.

However, the difference between leakage and exposure tests, in terms of the end action is important. A leakage standard is basically a pass–fail standard so the likely outcome of a test failure is that work has to be done to reduce the leakage. After remedial action is completed, regular leakage check tests are then instituted.

With intentional radiation there is normally no desire to reduce the radiation level since the power used is that required for the operational function. Thus the task is rather to identify potential hazards and then keep people away from them. This means that there is usually a continuing requirement for active safety management in situations where there may be day-to-day changes in the hazard situation. This is a particular problem on test sites where new equipment is tested before despatch to customers, because of the regular movement of equipments in an on-going activity.

Exposure standards deal with the limits for human exposure to RF fields over the range of frequencies used or likely to be used. The situations in which exposure standards apply will be those where there is intended radiation which can interact with people and with the environment by way of reflections of energy from structures and the ground.

Such reflection can considerably enhance the fields to which people are subjected. Of course, there will be cancellations from out-of-phase reflections as well, but the safety management aspect requires concentration on enhancements which increase exposure, where these occur. The magnitude of possible field enhancements should not be underestimated as the increase can be as much as four times or more. In power terms, this is 6 dB or more.

In particular, where RF is reflected from a resonant antenna or anything which fortuitously acts as one, there can be a concentration of the field with increases in excess of the figure mentioned above. This is particularly the case at microwave frequencies where the operation of antennas is quasi-optical. This problem is a common one on a development site where unused antennas are left in the area.

In practice, there will often be more than one RF source at a given location and transmitting stations, in particular, often involve an appreciable number of transmitters. This can lead to the possible multiple irradiation of people employed outside in the proximity of antennas.

There can still be cases where power might have to be reduced when either the radiation is in the public domain and in excess of permitted levels or where personnel in an occupational situation cannot be excluded from the field because they have a task to undertake in a place where the radiation level is too high.

It may also be necessary where there is a risk to flammable substances or EEDs.

In the latter cases, the substances or devices may be on the same site as the transmitters or they might be nearby in the public domain and not owned or controlled by the transmitter operator. There may even be cases where the siting of a transmitter or transmitters cannot be contemplated at a particular place because examination of the proposed site and the neighbouring installations and facilities owned by other people, indicates that the risk is too great.

General features of standards

Basis

A common feature of all the standards discussed here is that a specific absorption rate (SAR) of $0.4\,\mathrm{Wkg^{-1}}$ is specified as the basis of occupational standards (see Chapter 3). They also make provisions for higher SAR values in a very limited mass of tissue (e.g. $1\,\mathrm{g}$ or $10\,\mathrm{g}$).

The choice of $0.4\,\mathrm{Wkg^{-1}}$ is based on thermal considerations and is considered to be one order below that which gives rise to observable effects. SAR is, of course, not measurable for a living person and cannot be used for practical work.

For theoretical purposes, measurements can be made with dummies as described earlier or modelled on a computer. Both suffer the limitation of not providing a simulation which reproduces accurately the thermo-regulatory capability of the human body.

Across the range of frequency used for RF radiation, the basis of safety standards does, increasingly, reflect the nature of the RF interaction with the human body as new information is obtained about these interactions. For example, at the low frequency end of the RF spectrum, induced body and limb current densities are now the limiting factors. In the HF/VHF region human whole-body resonance occurs as a function of height and hence provision has to be made in standards to ensure that the SAR is not too high.

At the highest end of the RF spectrum in the gigahertz region, interest is centred on the deposition of energy in a small skin depth, i.e., the superficial skin layers and the possibility of the overheating of the skin. Thus most standards require people to get out of any microwave field which gives them a sensation of warmth in the skin, whether or not the power densities exceed permitted values.

The various standards discussed in this chapter, spreading as they do over about ten years, reflect these various aspects in the light of the knowledge available when they were written. The most recent standards obviously provide the greater insight into current views and beliefs on human exposure to RF.

In order to determine measurable safety limits, it is necessary to define derived limits, in terms of numeric limits for power density and the electric and magnetic field components which will ensure that the specified SAR is not exceeded. In addition, at the lower frequencies induced body and limb current limits now have to be specified.

The 'occupational' and 'public' limits issue

Some bodies concerned with standards consider that the limits for these two categories should be different, those for the public being lower than those for the occupational category. The medical and other aspects of the arguments relating to the two approaches are discussed in Chapter 3.

In the limit, these amount on the one hand, to those who believe there is some scientific basis for provision of additional protection for the public and and therefore advocate reduced levels. On the other hand, there are those who do not see any basic technical reasons and consider the matter as a social and political one.

It will be seen later in this chapter that the trend seems to be towards providing separate limits for the public and for occupational purposes. As noted previously, it is necessary to define the two categories since many people in an organisation which uses RF sources may not be technical people. Such people are generally treated as 'the public' in the sense that they have not chosen to work with RF and are not knowledgeable on the subject. This leads to a need to control access to areas where fields might exceed the limits for the public even though such areas lie within premises owned or controlled by the employer.

All standards cover both cases, but the applicable safety levels may or may not be identical. Where separate lower limits are specified for the public, they are typically one fifth of the occupational power density limits with corresponding division by $\sqrt{5}$ for the electric and magnetic fields.

Contact currents

When metal masses are situated in an RF field, particularly at the lower frequencies it has always been known that it is possible to incur shocks and burns from contact with the masses. Such masses may be scrap metal, stored antenna mast sections and similar objects or even motor vehicles. The likely contact currents are not easily calculable for undefined masses of material, although some calculations are referenced in Chapter 3.

The consequences, in terms of possible burns, are equally difficult to forecast since the contact area of say, the hand, touching the metal mass determines whether the subject is burned or not. A large contact area such as a firm grip gives the least chance of burn compared with a point contact. However, few safety officers would be happy to suggest that anything remotely electrical should be grasped firmly! In any case, most such contacts, at least for those aware of the potential hazards, are probably accidental from stumbling or stepping back and likely to involve an unpredictable contact area.

Limb currents

The relatively recent research into currents induced in the human body when exposed to low frequency fields, has led to the recognition of such currents as being very significant below perhaps 100 MHz. The problem is current density (current per unit area) which is not a measurable parameter for human beings.

Consequently, any practical limits set are in terms of current rather than current density, making assumptions about the cross-sectional areas associated with the current flow. The basis of the concern about excessive current density in limbs, is that high localised SARs can result.

For the purposes of defining limb currents in a way that could be measured, there is usually a formula for the maximum permitted current in the limbs.

The measurement of current flow in limbs is mentioned in Chapter 6. However it cannot be described as a fully established measurement technique at this juncture. Three out of the four standards discussed in this chapter cover the issue of contact currents but only two deal with induced currents in limbs.

Standards generally attempt to specify the low frequency field limits in such a way that the likelihood of excessive limb currents occurring is minimised. Sometimes the levels concerned will not necessarily be the normal limits, a reduction below the normal field limits being recommended or implied in some standards because of possible induced current levels.

Nevertheless it may be the case that in some situations these currents will still have to be measured because our knowledge of the relationships between field levels and limb currents is not adequately defined. Once good current measurement practices are established and put into use, better evidence about limb currents will become available.

Time averaging

Since about 1960, one-tenth hour (6 minutes) averaging has been established and limits in standards are specified as the average power densities or other field quantities over a period of 6 minutes.

Consider, for example, a continuous limit for power density of $50\,\mathrm{Wm}^{-2}$. This, over 6 minutes, can be expressed as an energy density in watt–hours or joules. In watt–hours this is:

$$50\,\mathrm{Wm}^{-2} \times 0.1\,\mathrm{h} = 5\,\mathrm{Whm}^{-2}.$$

Hence an exposure to $100\,\mathrm{Wm}^{-2}$ for three minutes (0.05 of an hour), followed by zero exposure for the next three minutes would also give:

$$100\,\mathrm{Wm}^{-2} \times 0.05 = 5\,\mathrm{Whm}^{-2}.$$

In joules the permitted limit would be $50\,\mathrm{Wm}^{-2} \times 360$ seconds $= 18\,000\,\mathrm{Jm}^{-2}$.

Clearly, some care must be used in the maximum power density allowed in practice, lest some instantaneous damage should occur. Six-minute time averaging is the basis for the use of the rotational factor for constantly rotating beams discussed in Chapter 5 and for other forms of time-averaging in Chapter 9.

Peak pulse energy

There has always been some concern about the possible effects of exposure to the very high peak pulse powers used on modern radars and other pulsed equipment.

Whilst there is no specific evidence of adverse effects on people, the very high values of power density involved inevitably cause most people to feel a little concerned. Again three of the four standards have some form of limit for the energy in the pulse. Limits set for peak pulse energy are usually given in joules per square metre (Jm^{-2}) or in terms of peak pulse power density. A basic calculation is given in Chapter 5. In some cases the form of the calculation is given in the standard and needs to be followed explicitly.

Progress in RF safety standards

The general UK and USA standard over thirty years or so ago was a uniform limit of $100\,\mathrm{Wm}^{-2}$ across what seems now to be a very limited frequency range, 30 MHz to 30 GHz in the case of the UK. It made no distinction between 'occupational' and 'public' exposures.

In the UK general guidance was published by the Home Office in 1960[26]. This gave some useful general guidance regarding rotating beams and about the avoidance of risks to the public but promulgated the flat 100 Wm-2 standard for all purposes. The document is historically interesting but now obsolete.

In the USA, ANSI produced a standard in 1974 and again in 1982. In the latter, important changes were beginning to take place. The 1982 ANSI standard, ANSI C95-1.82[19] was a more detailed standard which, among other things, recognised the fact that the SAR increases as the frequency approaches the resonant frequency of human beings.

As noted earlier, the resonant frequency of an erect person will depend on height. A plot of SAR versus frequency will have a shape such that it reaches maximum over the range of resonance frequencies for children and adults.

An interesting chart in ANSI 1982 shows how the SAR-frequency relationship was established using 15 sets of research results from various investigators. The single set of limits in this standard is applicable to occupational and public use without distinction. The standard covers the range 0.3 MHz to 100 GHz, indicating the movement of microwave development into the higher gigahertz region.

In 1984, the INIRC committee of IRPA published recommendations and in 1988 further recommendations. INIRC 1988[27] went much further than ANSI 1982 in that it had separate, lower, limits for the exposure of the public based on a SAR of $0.08\,\mathrm{Wkg}^{-1}$ (one fifth of the occupational basis). It also dealt with peak pulse energy, and body contact current limitation.

In 1989 the UK NRPB issued document GS11[28] giving guidance on safety with electromagnetic fields and radiation. Generally speaking, the reference values, as the numerical values in the document are termed, are close to the occupational limits of INIRC 88 standard. The separate limits for the public were not adopted. Peak pulse energy, contact current and induced currents in the limbs are covered. The latter was new and stemmed from the work of Gandhi and others referred to earlier.

The NRPB is currently engaged in revising GS11 with a timescale which suggests publication in 1993.

In 1992 the former ANSI committee, now operating as part of the American IEEE, published standard IEEE C95.1-1991 which replaces ANSI 82. This is therefore the most recent standard to be published[4].

Comparison of standards

It is useful to compare the ANSI 1982, INIRC 1988 and the NRPB 1989 standards first and later to examine the IEEE 1991 standard because of the considerable development of the latter relative to earlier standards. For simplicity in referring to standards, the short form identities are used e.g. ANSI 82, the digits being the last two digits of the year of issue. Also, not withstanding the fact that technically some of the documents are standards and some guides, the term 'standard' is used for them all.

There is a general use of the title radio frequency protection guide (RFPG) in the earlier ANSI standards, whereas elsewhere a variety of terms are used such as limit value, reference value, exposure limit, permitted exposure limit (PEL) and maximum permissible limit (MPE). In this chapter the word 'limit' is used to cover all these terms and is qualified only where thought essential.

Occupational and public limits

Table 4.1 shows the status of the standards. It can be seen that only the INIRC 88 standard provides different limits for public exposure.

NRPB 89 has one provision, not identified in Table 4.1, for low RF frequencies whereby, if the public are put at risk of incurring burns from conductive objects in the field in the public domain, then a reduction of the permitted limits is specified. There is a similar provision within INIRC 88.

Table 4.1   *Standards and public exposures*

Standard	Occupational limits	Public limits
ANSI C95.1–1982	Yes	Same as occupational
NRPB GS11 1989	Yes	Same as occupational
INIRC 1988	Yes	Yes, separate (lower) limits

Power density

The power density limits of ANSI 82 are shown graphically in Figure 4.1 to illustrate the basic shape of the limits characteristic. Figure 4.2 shows the three standards plotted together, but starting at 30 MHz to provide good resolution. The problem here is that the NRPB 89 and INIRC 88 start

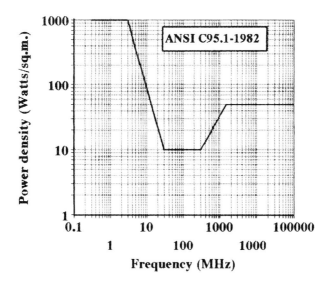

Figure 4.1 *ANSI C95. 1–82 standard power density limits*

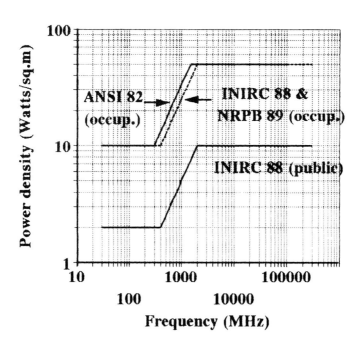

Figure 4.2 *Comparison of three standards: limit values for power density*

specifying power density at 30 MHz and 10 MHz respectively, requiring measurements below those frequencies to be made for the electric and magnetic fields separately. On the other hand, ANSI 82 specifies the equivalent plane wave power density down to 300 kHz.

It can be seen from the graph of Figure 4.1 that there is a marked trough in the shape of the curve in recognition of the resonance of human beings at frequencies determined by their height. Since the phenomenon of resonance results in a maximum absorption of RF energy, it follows that the limit set by an exposure standard should fall to its lowest level (i.e., tightest limit) where the SAR is maximum, so that the SAR at resonance is held at a safe level.

As a result, the graphical shape of the three standards takes the form of what is often referred to as a negative-going pulse shape, the two edges being in the form of ramps. The frequency range across the bottom of the trough is from 30 MHz to 300 or 400 MHz, according to the standard being considered, as shown in Figure 4.2.

Theoretically, for occupational purposes, this frequency range will be more constrained than for the public generally since it is determined by the range of heights of adults, whereas the 'public' includes children and will provide a much wider range of potential resonances ranging from adults to very small children. However, in practice, the width of the trough is made such as to cover both groups and does not change in those standards which separate occupational and public limits.

It will be noted that the occupational limits of the three standards do not differ significantly except with regard to the ramp up into the gigahertz region where the ANSI 82 standard differs from the other two by 100 MHz at the bottom of the ramp and 500 MHz at the top. Table 4.2 lists the full

Table 4.2   *Comparison of three standards – power density*

ANSI 82		INIRC 88 (occupational only)		NRPB 89	
Frequency	$mWcm^{-2}$*	Frequency	Wm^{-2}	Frequency	Wm^{-2}
0.3–3 MHz	100	0.1–1 MHz	–	0.003–3 MHz	–
3–30 MHz	$900/f^2$ (MHz)	1–10 MHz	–	3–30 MHz	–
30–300 MHz	1	10–400 MHz	10	30–400 MHz	10
0.3–1.5 GHz	$f/300$ (MHz)	0.4–2 GHz	$f/40$ (MHz)	0.4–2 GHz	25f (GHz)
1.5–100 GHz	5	2–300 GHz	50	2–300 GHz	50

Note: *All values are in the original units.

numeric values given in the standards. In the graphs, the ANSI 82 limits have been converted from $mWcm^{-2}$ to Wm^{-2} for plotting purposes.

Electric and magnetic field limits

Figure 4.3 shows the electric field limits for the three standards, starting at 300 kHz. Here again the start points for the standards differ, the NRPB 89

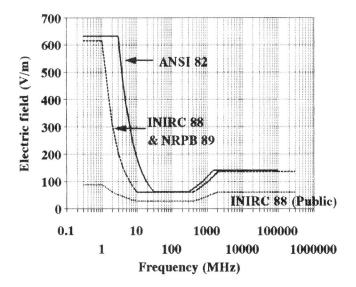

Figure 4.3 *Comparison of three standards: limit values for the electric field*

standard starting at 3 kHz, the INIRC 88 standard at 100 kHz and the ANSI 82 standard from 300 kHz. As before, the INIRC and NRPB limits are virtually identical and the ANSI 82 standard has greater limits over the ramp down to 30 MHz, after which the difference is on the ramp up to 1.5 GHz. In the graph, the ANSI 82 limits have been converted from $(Vm^{-1})^2$ to Vm^{-1} for plotting purposes. The values are given in Table 4.3 in the form given in the standard.

The corresponding occupational magnetic field limits are shown in Figure 4.4. The graph starts at 3 MHz to give reasonable resolution as the permitted limits rise rapidly below that frequency. Apart from the 3 MHz end of the INIRC curve, there is little difference between the standards over most of the frequency band. The NRPB 89 and ANSI 82 curves virtually lie on top of each other. The numerical values can also be found in Table 4.3.

Table 4.3   *Comparison of three standards – electric and magnetic fields*

ANSI 82		INIRC 88 (occupational only)		NRPB 89	
Frequency	$(Vm^{-1})^2$ $(Am^{-1})^2$	Frequency	Vm^{-1} Am^{-1}	Frequency	Vm^{-1} Am^{-1}
MHz		MHz		MHz	
0.3–3	400 000	0.1–1	614	0.03–1	614
	2.5		1.6/f	(see note 1)	4.89/f
3–30	4000(900/f²)	1–10	614/f	1–10	614/f
	0.025(900/f²)		1.6/f		4.89/f
30–300	4000	10–400	61	10–30	61.4
	0.025		0.16		4.89/f
GHz		GHz			
0.3–1.5	4000(f/300)	0.4–2 GHz	$3\sqrt{f}$	30–400	61.4
	0.025(f/300)		$0.008\sqrt{f}$		0.163
				GHz	
1.5–100	20 000	2–300 GHz	137	0.4–2	$97.1\sqrt{F}$
	0.125		0.36		$0.258\sqrt{F}$ (see note 2)
				2–300	137
					0.364

Notes:
1   This standard gives field limits down to 'f <100 Hz', but only 30 kHz upwards is given here.
2   F = GHz ; lower case f = MHz everywhere.

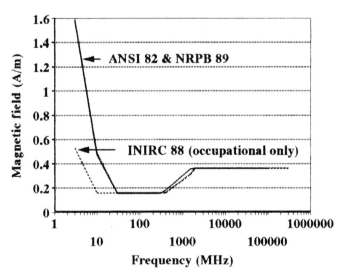

Figure 4.4   *Comparison of three standards: limit values for the magnetic field*

Other aspects

Of more interest is the fact that the later two standards define a limit on contact current resulting from a physical contact with a conductive object in the field of an antenna, in order to prevent burns resulting. NRPB 89 also introduces limits on body and limb currents as a key factor in the determination of human safety at the lower frequencies (less than 100 MHz).

Both INIRC 88 and NRPB 89 specify peak pulse density or energy for pulse transmission. Table 4.4 shows, in abridged form, these aspects of the standard – limb currents, contact current, peak pulse power density/energy and SAR values.

Table 4.4   *Standards comparison – other limits data*

Standard	Induced current	Contact current	Peak pulse energy/ p. density
ANSI 82			
0.4 Wg^{-1} basis	N/A	N/A	N/A
INIRC 88	Not specified	Limit 50 mA	1000 × p.d. limits;
Occupational			32 × field strength
0.4 Wkg^{-1} basis			limits
Public 0.08 Wkg^{-1}	Not specified	Limit 50 mA	As above but using
basis			the public limits
NRPB 89 0.4 Wkg^{-1}	Maximum 1+f/1500	See previous	For pulses < 50 µs
basis	or 100mA (lesser)	column	limit is 0.4 Jm^{-2}
	f=Hz		

Recent developments – the IEEE C95.1-1991 standard

In 1990 the ANSI C95 committee of the IEEE produced the 'final final' draft of a revision of the ANSI 82 standard. This standard is now published as the IEEE C95.1–1991 standard. The standard is worth studying in some detail since it may be thought to point the way to the future.

The material here relating to ANSI C95.1-82 and IEEE C95.1-1991 is reproduced by kind permission of the American Institution of Electrical and Electronic Engineers. Anyone proposing to use the standard should study the full document because of the number of qualifications therein.

Before discussing the contents, it is worth noting that power density is still given in mWcm^{-2}, rather than in Wm^{-2}. However, the electric and magnetic field values are now given directly rather than in the form of the squares of the field values, as was the case in ANSI 82. The squares of the field units were quite unpopular as an instrument scaling because of the need to use a calculator when comparing with other standards which all used direct values (Vm^{-1} and Am^{-1}).

Such manipulations coupled with the further need to convert power density limits in $mWcm^{-2}$ into Wm^{-2} when comparing with other standards, provide another risk of error.

Occupational and public limits

The most profound change is the acceptance of the need to provide for separate and, in principle, different limits for occupational purposes and for the public, though these terms are not used in the standard. Instead, the concept of a 'controlled environment' is used for occupational purposes and an 'uncontrolled environment' for those not occupationally engaged with RF generating equipment.

There are provisions in the standard to accept the transient movement of people who are not occupationally engaged, in and out of controlled areas, presumably to accommodate secretarial and other support staff.

Whilst the applicability of controlled and uncontrolled environments is easy to understand in the context of RF transmission, the problem of separating people from RF radiation will also arise, at least in principle, for RF process machines, medical therapy machines and other sources.

In practice, in these cases, the problems may be circumscribed by the limited range of stray radiation and leakage in such equipment since the design of RF machines attempts to confine the fields to their function e.g. heating a work piece with an RF machine, whereas transmitter antennas are intended to transmit energy to distant places.

Power density limits

Most standards now limit the applicability of power density measurement below some specific low frequency limit e.g. NRPB 89 specifies 30 MHz and INIRC 88 specifies 10 MHz. Measurement below these frequencies involves the separate measurement of the electric field and magnetic field against specific limits for each.

IEEE 91 specifies power density from 100 MHz upwards, relegating the equivalent plane wave power densities for other frequencies to a 'for information' status. Figure 4.5 shows the power density limits for both environments. Two points of interest may become obvious immediately when the standard is compared with the corresponding provisions of the ANSI 82 standard:

1 The upper limit for occupational purposes at the high end of the frequency range is now $100\ Wm^{-2}$ instead of $50\ Wm^{-2}$.
2 The 'uncontrolled environment' limit in that region is the same i.e. there is no distinction between the two groups, whereas at the opposite end of the spectrum the uncontrolled limit is one fifth of the controlled limit. The numerical values are given in Table 4.5.
3 The ramp from 300 MHz upwards is to 3 GHz instead of 1.5 GHz.

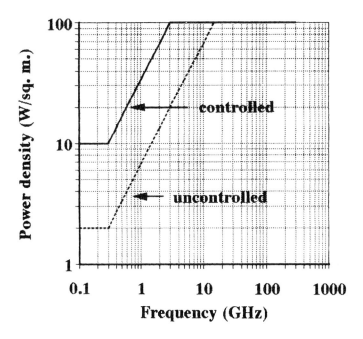

Figure 4.5 *IEEE C95. 1–1991 standard: power density limits for both environments*

Table 4.5 *IEEE C95.1-1991 Power density limits*

Controlled environment (Wm-2)	Frequency (MHz)	Uncontrolled environment (Wm-2)
10	100 to 300	2
f/30	300 to 3 000	f/150
100	3 000 to 15 000	f/150
100	15 000 to 300 000	100

Notes:
1 f is in megahertz
2 Power density figures and expressions have been changed to give Wm^{-2}.

Electric and magnetic field limits

Figure 4.6 compares the electric field limits and Figure 4.7 the magnetic field limits. Table 4.6 gives the limits in numerical form. For controlled

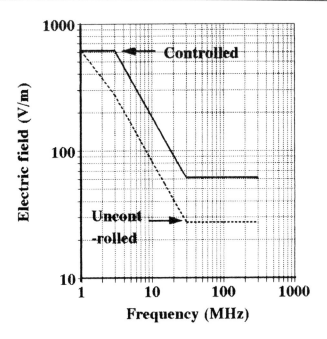

Figure 4.6 *IEEE C95. 1–1991 standard: electric field limits for both environments*

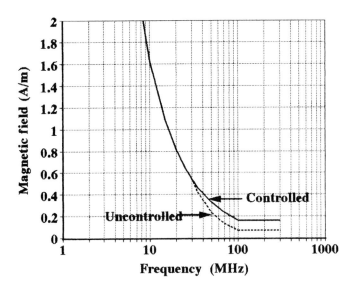

Figure 4.7 *IEEE C95. 1–1991 standard: magnetic field limits for both environments*

Table 4.6 *IEEE C95.1–1991 Electric and magnetic field limits*

Controlled environment		Frequency (MHz)	Uncontrolled environment	
Vm^{-1}	Am^{-1}		Vm^{-1}	Am^{-1}
614	163	0.003 to 0.1	614	163
614	16.3/f	0.1 to 1.34	614	16.3/f
614	16.3/f	1.34 to 3	823.8/f	16.3/f
1842/f	16.3/f	3 to 30	823.8/f	16.3/f
61.4	16.3/f	30 to 100	27.5	$158.3/f^{1.668}$
61.4	0.163	100 to 300	27.5	0.0729

NOTES:
1 f is in megahertz
2 No values specified above 300 MHz

environments, the highest electric field limits (at the low frequency end) are a few percent lower than for the ANSI 82 standard ($614\,Vm^{-1}$ against $632\,Vm^{-1}$). The controlled and uncontrolled limits start to be differentiated above 1.34 MHz, so that the difference is a factor of $\sqrt{5}$ above 3 MHz.

The magnetic field limits have been eased somewhat at the low frequency end of the controlled environment limits. The uncontrolled environment limits are the same up to 30 MHz where they then reduce until at 100 MHz they are less by the expected factor of $\sqrt{5}$.

Partial body exposures

Where whole body exposure is not uniform, a common case for low power sources and other situations where only a part of the body is affected, compliance with the limits in the standard is established by the spatial-averaging of the power density or the mean squared electric and magnetic field strengths. This is averaged over an area equivalent to the vertical cross-section of the body.

In order to limit the highest values of exposure included in the averaging operation, these are defined by tabulated values which, for frequencies up to 300 MHz are $< 20\,(E^2)$ or $< 20\,(H^2)$, where E and H are the electric field and magnetic field spatial average values from the tables of limits for controlled and uncontrolled environments (Table 4.6).

For power density above 300 MHz, the maximum values range from $< 200\,Wm^{-2}$ up to 6 GHz and $400\,Wm^{-2}$ from 96 GHz onwards for controlled areas. For uncontrolled areas, the corresponding limits for these two cases are $40\,Wm^{-2}$ and $200\,Wm^{-2}$ respectively.

The provisions for partial body exposure exclude the eyes and the testes.

Limb and contact currents

Table 4.7 gives the limits for limb currents (one foot and two feet) and contact currents up to 100 MHz. The limit for two feet is double that tabulated for a single foot. Currents are averaged over 1 second. Figure 4.8 illustrates contact and foot current (one foot) over the frequency range to 100 MHz.

Table 4.7   *IEEE C95.1–1991 Limb and contact currents*

Controlled environment Each foot and contact current (mA)	Frequency (MHz)	Uncontrolled environment Each foot and contact current (mA)
1 000f	0.003 to 0.1	450f
100	0.1 to 100	45

NOTES:
1 f is in megahertz
2 Limits for both feet are double those above.

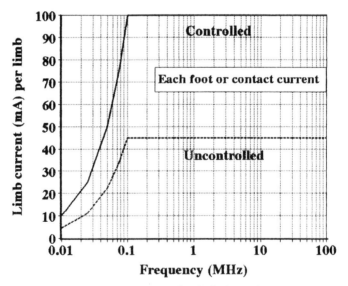

Figure 4.8   *IEEE C95. 1–1991 standard: limb and contact currents*

Time averaging

There are a number of different provisions for the time averaging of quantities. These are summarised below:

Controlled environments

The familiar 6 minutes averaging time is used up to 15 GHz for all the measured quantities other than limb and contact currents. Above 15 GHz, the time reduces progressively, as shown in Figure 4.9. The intention is to protect against the possibility of burns at the higher microwave frequencies where energy is largely deposited in the superficial layers of the skin.

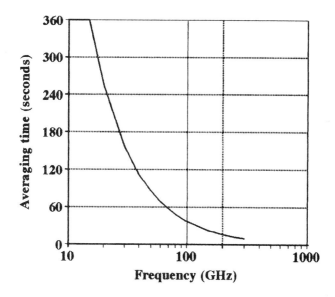

Figure 4.9 *IEEE C95. 1–1991 standard: time averaging above 15 GHz*

Uncontrolled environments

The provisions here start with the usual 6 minutes up to 1.34 MHz. There are then several changes before 15 GHz where the same arrangement as for the controlled environment applies. The variations are described under two headings:

1 For power density (100 MHz upwards) and the electric field (up to 300 MHz) there are two additional requirements. At frequencies of 1.34 to 3 MHz, the averaging time is according to an expression which results in the averaging time spread of 6 minutes at 1.34 MHz to 30 minutes at 3 MHz and this time period then continues to 3 GHz.

From 3 GHz to 15 GHz, the expression for averaging time brings the time period back from 30 minutes at 3 GHz to 6 minutes at 15 GHz.

2 For the magnetic field, the averaging time of 6 minutes continues to 30 MHz. From 30 to 100 MHz, the expression used takes the time period up to 30 minutes at 100 MHz and this continues to 300 MHz. Magnetic and electric field limits are not given above 300 MHz since power density is used thereafter.

Pulsed radiation

The general ceiling provision for peak pulse electric field is 100 kV^{-1} from 0.1 to 300 000 MHz. For pulse durations less than 100 ms and a maximum of 5 pulses in 100 ms, an expression is given for peak pulse power density as follows:

$$\text{Limit } S_{pk} = \frac{\text{MPE} \times \text{averaging time (seconds)}}{5 \times t_p \text{ (seconds)}}$$

Where MPE is the E field equivalent power density from the limit table for controlled or uncontrolled environments, as appropriate, S_{pk} is the peak pulse power density and t_p is the pulse dilution.

If there are more than 5 pulses in 100 ms or the pulse duration exceeds 100 ms the energy density in 100 ms is limited by the following expression:

$$\Sigma S_{pk} \times t_p \text{ (seconds)} = \frac{\text{MPE} \times \text{averaging time (seconds)}}{5}$$

Exclusions

There is a considerable amount of text relating to exclusion clauses. These are not covered here. However the low power source exclusion is worth mentioning because it changes the similar provision in its predecessor.

In controlled environments

The user of a source of 7 W or less may exceed the permitted limits between 100 kHz and 450 MHz. At frequencies between 450 MHz and 1500 MHz, the power should not exceed:

$$P = 7 \times (450/f) \text{ W}$$

Where f is the frequency in MHz. These exclusions do not apply to radiating structures within 2.5 cm of the human body.

In uncontrolled environments

The exclusion relates to devices which emit RF energy without the control or knowledge of the user. At frequencies between 100 kHz and 450 MHz the permitted limit may be exceeded if the power does not exceed 1.4 W.

At frequencies between 450 MHz and 1500 MHz, the power should not exceed:

$$P = 1.4 \times (450/f) \text{ W}$$

Where f is the frequency in MHz. These exclusions do not apply to radiating structures within 2.5 cm of the human body.

Static (DC) magnetic fields

Static magnetic fields have little to do with the technical subject matter of this book. However, in the course of doing RF radiation measurements on some types of equipment, particularly in the medical field, such magnetic fields may be encountered.

Most of the standards reviewed above do not specifically deal with this subject but the NRPB has produced separate guidance on it in the form of a draft amendment to NRPB GS11. This specified a whole-body limit of 200 mT for an 8 hour period. A later NRPB report[29] reviewed the subject and suggested that the average exposure over one day for occupational exposure should not exceed 200 mT. The general concerns are:

1 Effects on the human body.
2 'Flying objects' due to ferrous materials brought into the vicinity, being accelerated by the field to hazardous velocities.
3 Adverse effects on instruments containing components susceptible to damage from the field.

Summary

Over about twenty years the standards and guides produced by expert bodies have developed in the extent of their coverage and in the refinement of limits, as new research findings have become available.

The IEEE C95.1–1991 is the latest standard produced and this makes a major departure from its predecessor. It establishes a system of dual limits, choosing to relate these to controlled and uncontrolled environments instead of the usual 'public' and 'occupational' situations.

It provides some complications in practical use, in that averaging times vary with frequency. This will affect instruments which offer time averaging with a fixed averaging time, though this mainly applies to the uncontrolled environments. However, instruments will no doubt become programmable in the fullness of time.

There is no doubt that there is a need for an international standard so that instrument producers and users do not have to face a variety of different standards. This is particularly important in Europe where the EEC is currently drafting limits for RF radiation, and this will presumably also be followed by the EFTA nations, whereas virtually all of the RF radiation instrument manufacturers are in the USA!

Differences in standards can sometimes make it difficult for instrument designers to satisfy all parties adequately without producing special versions.

The various exclusions in the standard do deal with some of the minor apparent infringements of standards which cause a great deal of nuisance without giving rise to any known risk e.g. exposures of a small part of the body to levels a little in excess of the standard such as can occur with man-pack radio sets, and the like.

Reflections on the use of standards

Any views on the use of standards are likely to be personal views, and the following are no exception. However they do stem from long experience and may be of some value:

1 Do not work at levels right up to the permitted limits when there is no need to do so. Many people are inclined to set safety boundaries at such limits routinely when no one needs such a tight boundary.
2 Similarly, do not use rotating beam rotational averaging to set limits if no one needs to work that close.
3 Do not use exclusion clauses as a substitute for dealing with a problem, if it is soluble.
4 The electric field values in standards do not necessarily guarantee that limb and contact currents will not be exceeded. Caution is needed in cases where exposures are close to the limits and measurements may be necessary.

In summary, this means relating measurements and boundaries to real requirements. Each time a new standard is issued which tightens up limits in some way, we have people wondering why they were previously allowed to be exposed to levels of RF energy which are now forbidden. Practical experience shows that this need not always be the case if efforts are made to minimise exposure rather than to just accept the limit values given in standards.

5

Antenna system and other basic safety calculations

Antenna survey information requirements

Unless some form of calculation is available to provide an idea of the characteristics of an antenna, a safety survey becomes a journey into the unknown in which the surveyor may be put at risk and may also do a great deal of unnecessary work. Antenna radiation surveys are essentially an exploration of a volume since the work and the public environments have a height dimension as well as the obvious area dimension.

Calculations can be undertaken to great depths using computers, with a corresponding magnitude of expenditure. Some of this may be necessary if the specification for the item demands such treatment, but this will then apply primarily to the designer and the supplier in proving compliance with the performance specification, rather than the safety aspects. Of course, having done such calculations they may also serve for the latter purpose.

For the equipment user and those not having available the more exacting calculations done by the designer, it is possible for more limited calculations to be done quickly and cheaply with standard formulae, particularly in the microwave antenna field, if the limitations of each formula are accepted.

In any case, theoretical calculations will provide information about an antenna in free space, an idealised situation.

In practical working situations the environment may appreciably affect the measured radiation levels, the flat site having most effect when the beam is set at negative elevations and ground enhancements from reflections arise. More common environments have buildings, metal structures and vehicles and the nature and location of some objects may vary from day to day due to the other activities being undertaken. Few sites are without such buildings and structures where there may be interaction with whatever objects are around and the ground may often be far from flat.

It is clear that the calculations required for surveys only need to give an approximation of the power flux densities or other field quantities involved so that the surveyor can determine the order of things to be expected. Insofar as calculations normally reflect the hypothetical free space condition, there is no value in having such calculations done to great accuracy when a factor of 4 to 6 dB may apply to reflection enhancements from the ground or from structures. The value of calculations is firstly to ensure the safety of the surveyor by highlighting areas of potentially high exposure and secondly to help in reducing the volume to be explored by the surveyor and to confirm the order of the levels expected to be found in measurements.

Microwave calculations

General

Microwave transmission commonly uses a parabolic or paraboloidal reflector with a rectangular, circular or elliptical aperture. It is necessary to have some method of calculating the power densities likely to be associated with such antennas at the frequencies and power levels used.

A system[30] devised by a colleague of the author, Dr D. H. Shinn, formerly of the GEC-Marconi Research Centre, provides a relatively simple method of producing calculations for most communications and radar antenna systems, other than very specialised types such as those used to communicate with satellites. It uses three normalised contour diagrams on which can be plotted a reference value. The values of the individual contours are then easily obtained from the scaling on the charts.

The horizontal axis is scaled in multiples of the Rayleigh distance and the vertical axis is scaled in terms of the relevant antenna dimensions e.g. circular antenna reflector diameter. The system has been used by the author for many years and found very useful for the purposes discussed above. Because of the area of work in which Dr Shinn was involved, there is radar terminology used in the original paper although a communication system example is given here since the method is not specific to the radar field.

The term 'Rayleigh distance' may require some explanation. Figure 5.1 illustrates the concept involved. The Rayleigh distance has no standard definition and therefore needs to be specifically defined. The definition used here is the original Lord Rayleigh one ($D^2/2\lambda$), where, in the case of the circular aperture antenna, D is the diameter and lambda the wavelength. In the figure, this corresponds to the distance PN when the distance PM is greater than PN by $\lambda/4$. It follows that the phase of the radiation reaching P from M is delayed by 90° relative to that from point N.

Sometimes Rayleigh distance is defined as D^2/λ (45°) or $2D^2/\lambda$ (22.5°). The value in using a definition which gives the shortest distance which can still give reasonably accurate on-axis power density estimates, is the fact that the region adjacent to the antenna and the early part of the beam is frequently

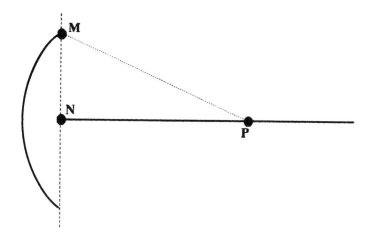

Figure 5.1   *Rayleigh distance illustration*

the one which is most likely to be accessed by personnel and having calculations available for relatively short distances from the antenna increases confidence in doing surveys; $2D^2/\lambda$ gives a value for PN which is four times greater.

Figure 5.2 illustrates the Rayleigh distance versus frequency for three circular aperture diameters.

It is generally accepted that for an antenna radiating into free space, four distinct regions can be identified where the behaviour of the electromagnetic field from the antenna displays specific characteristics. These are:

1 The non-radiating reactive near field region (induction field) with a very short range of a fraction of a wavelength. This region is not usually of significance in the microwave spectrum, from the point of view of safety. It can, however, be important at lower frequencies.
2 The radiating near field region. Here the field varies considerably with distance from the antenna. In the microwave spectrum meaningful calculations are difficult. The usual conservative approach to safety is to use the expression for the highest possible amplitude of the power density as applying to the whole of this region, and then undertaking specific measurements.
3 The intermediate field zone.
4 The far field zone.

Figure 5.3 illustrates these field zones. The intermediate and far field zones are dealt with in this chapter. The calculation methods described here involve three normalised diagrams covering:

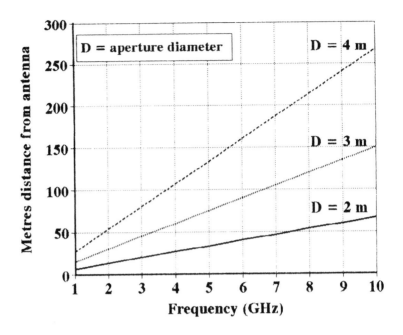

Figure 5.2 *Rayleigh distance versus frequency for three circular aperture diameters*

FIELD ZONES

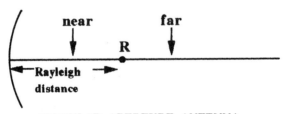

CIRCULAR APERTURE ANTENNA

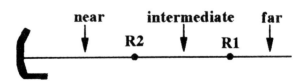

ELLIPTICAL/RECTANGULAR APERTURE ANTENNA

Figure 5.3 *Radiation field zones of antennas*

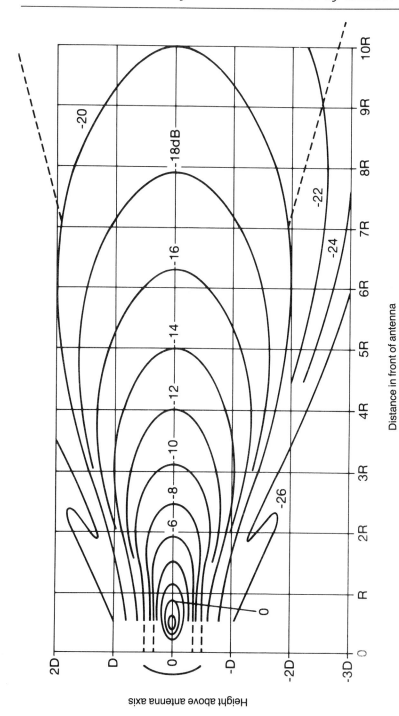

Figure 5.4   *Far field of circular aperture antennas*

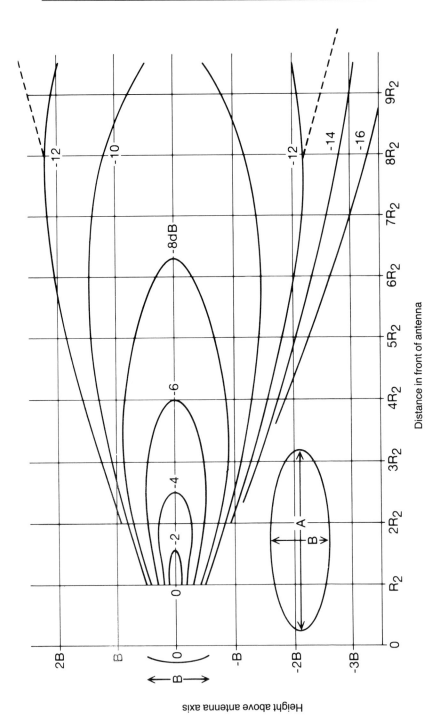

Figure 5.5 *Elliptical and rectangular aperture antennas–intermediate field with larger aperture dimension horizontal*

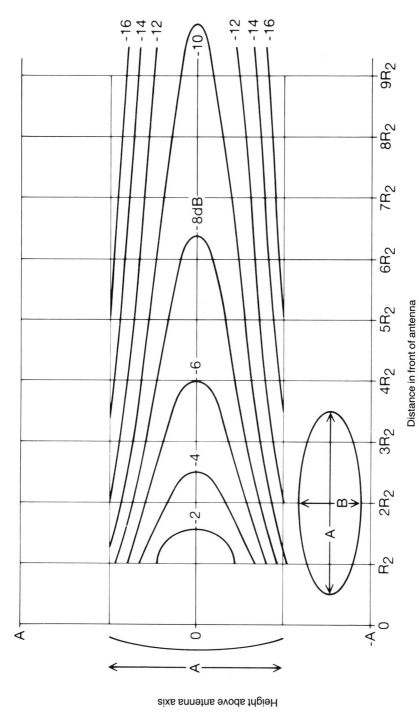

Figure 5.6 *Elliptical and rectangular aperture antennas–intermediate field with larger aperture dimension vertical*

Circular apertures: Far field calculation (Figure 5.4) (also used for the far field of elliptical and rectangular apertures).

Elliptical and rectangular apertures: intermediate field, vertical section; calculation for antennas with the larger aperture dimension horizontal e.g. a surveillance radar (Figure 5.5).

Elliptical and rectangular apertures: intermediate field, vertical section; calculations for antennas with their larger aperture dimension vertical e.g. a height finding radar (Figure 5.6).

Most of the calculations for elliptical apertures will apply to rectangular apertures, bearing in mind that no great precision is sought in these calculations. Note that the 'X' axis on all of these diagrams does not, except by coincidence represent the ground. A horizontal line can be drawn to represent the ground on the basis of the actual height of the antenna above the ground (see the example later).

Antennas with circular apertures

This is perhaps one of the most common types of microwave antenna consisting of a reflector of paraboloidal surface and circular boundary (almost universally known as a dish) with a waveguide or dipole feed at the focus. The first diagram, Figure 5.4, is used for determining the far field of these antennas. It will be noted that the vertical axis of Figure 5.4 gives height above and below the axis of the antenna in terms of units of 'D' the antenna diameter.

The horizontal axis is marked in values of R, the Rayleigh distance, the contours going down to 0.5R, below which the pattern can be complicated but there is likely to be little power flux beyond a distance of 0.6D from the axis in that region.

The field of this type of antenna consists of two regions, the near-field region between the antenna and R where the power is nearly all contained within a cylinder of radius 0.5D. In the far-field region from greater than distance R onwards, the power spreads out and the power density on the axis is given by the far field formula:

$$S_{FF} = P - T + G - 10 \log (4\pi r^2) \text{ dBWm}^{-2} \tag{1}$$

Where:

P = transmitter output power in dBW.
T = loss between transmitter output and antenna input in dB.
G = antenna gain in dBi, i.e., dB relative to an isotropic antenna.
r = distance in metres from the antenna at distances much greater than R.

Considering Figure 5.4 further, a reference value point is shown as zero on the diagram and subsequent contours are marked in two decibel steps down to −26 dB relative to the reference value. If the power density L at the reference point on the diagram is calculated in dBWm^{-2}, then the contour values are expressed by simple subtraction of their marked relative values. For example, if L = 20 dBWm^{-2}, then the first three contours will be 18, 16, and 14 dBWm^{-2} respectively.

Note that the first two contours are not marked but are the −2 dB and −4 dB steps. If the reference value is converted to Wm^{-2} then the other contours will need to be determined with the dB/ratio table, Table 5.1,

Table 5.1   *Conversion factors for antenna charts when plotting power densities*

Contour decibels	Ref. value multiplier	Contour decibels	Ref. value multiplier
−2	0.63	−16	0.025
−4	0.4	−18	0.016
−6	0.25	−20	0.01
−8	0.16	−22	0.0063
−10	0.1	−24	0.004
−12	0.063	−26	0.0025
−14	0.04		

Multiply reference value by the multiplier for the relevant contour.

multiplying the reference value in Wm^{-2} by the relevant factor from the table.

The expression for computing the reference value L is given below in equation 2. This equation is as equation 1 but with the Rayleigh distance (R) substituted for (r):

$$L = P - T + G - 10\log(4\pi R^2) \text{ dBWm}^{-2} \tag{2}$$

The following data is required:

P = mean power of the transmitter or peak power in dBW; see the note below on mean and peak pulse powers.

G = gain of antenna in dBi, measured at the input to the antenna

T = attenuation in dB between the transmitter output and the antenna input

D = diameter of the antenna in metres

λ = wavelength in metres (300/f MHz)

L = reference power flux at the point marked zero on the antenna axis of Figure 5.4. The actual power density on axis at the Rayleigh distance R is about −0.8 dB or 0.83 times the reference power flux at L.

R = Rayleigh distance (metres) defined as indicated earlier.

Where the gain data is not available, it can be estimated as follows:

$$G \text{ for} \simeq 64\% \text{ antenna efficiency} = 8 + 20\log(D/\lambda) \text{ dBi} \tag{3}$$

Where D and λ have the same meanings and dimensions as above.

Note that since calculations for pulsed transmissions will need both mean power density and peak pulse power density values, it can usually prove simpler to use mean power in such cases. The results are then plotted as Wm^{-2} and the peak pulse power density values computed by multiplying the mean power densities by the reciprocal of the duty factor. These can then be marked on the one diagram using, say, red ink to distinguish them.

Example 1

Assume CW system with the following characteristics

P = 1 kW (30 dBW); T = 0 dB; G = 40 dBi; D = 4 m; f = 3000 MHz; hence λ = 0.1 m.

The calculation steps are:

Step 1
Calculate $R = D^2/2\lambda$. The horizontal axis of Figure 5.4 can then be scaled. The vertical axis is scaled without calculation from the known value of D. The ground level can also be marked from a knowledge of the actual height from the antenna axis to ground level.

For example, if that height amounted to three times the diameter of the antenna then the horizontal axis of the diagram would correspond to ground level. Such positioning clearly depends on the size of antenna mount.

$R = 16/(2 \times 0.1) = 80 \text{ m}$.

Step 2
Calculate the reference value $L = P - T + G - 10\log(4\pi R^2)$
$L = 30 - 0 + 40 - 10\log(4 \times \pi \times 80 \times 80) = 70 - 49.05 \text{ dBWm}^{-2}$
$L = 20.95 \text{ dBWm}^{-2}$

Step 3
Not many people work in $dBWm^{-2}$ for this sort of work, but 20.95 could be entered in Figure 5.4 as the reference value and the values of the successive contours would reduce in 2 dB steps so that, for example the -6 contour value would be 14.95 $dBWm^{-2}$. Where the results are preferred in watts per square metre, this can be obtained with a further step.

Step 4
$S = \text{antilog } 20.95/10 = 124 \text{ Wm}^{-2}$.

In this case the reference level is 124 Wm^{-2} and the contour values can be obtained as shown in Table 5.2 by multiplying by the factors in Table 5.1.

Table 5.2   *Converted results of example 1 for contour plotting*

Contour	Wm^{-2}	Contour	Wm^{-2}
Reference	124	−14	5
−2	78.1	−16	3.1
−4	49.6	−18	2
−6	31	−20	1.24
−8	19.8	−22	0.78
−10	12.4	−24	0.5
−12	7.8	−26	0.31

In practice, before doing any of the steps of the calculation it can be useful to evaluate $2P/D^2$, where P is in watts and D in metres. This gives a rough idea of the value of L. In this case it would give:

$2000/16 = 125\ Wm^{-2}$.

This example is a CW communications one but if we suppose that the same data applied to the mean power of a pulsed system with a duty factor reciprocal of 1000, then the peak pulse power density can be obtained directly by multiplying the values in Table 5.2 by 1000. Such information is of value in evaluating compliance of the peak power densities against a provision in most RF standards which aims to restrict excessive peak exposures, and also in connection with hazards to flammable vapours and electrically-fired explosive devices.

In any particular practical case for both the calculation above and for those which follow, it may be necessary to take into account the fact that the axis of the antenna is not horizontal and that the ground in front of the antenna is not horizontal. For example, if the ground is level but the antenna is pointed upwards at an angle of 0.5°, then the ground, wherever it is located, can be represented by a straight line whose origin is at the point on the 'Y' axis (extended downwards if required) corresponding to ground level and tilted down at 0.5°. In the example above this would be 7 metres down at 10 R (800 m) from the antenna. This is illustrated in Figure 5.7 where, to reduce complexity, it is assumed that the ground level corresponds with the 'X' axis although as noted earlier this may not be the usual case. The antenna dimension used in the previous example (4 m) has been used in Figure 5.7, making the antenna centre 12 m above the ground.

Elliptical and rectangular apertures

The circular aperture antenna field was described in terms of a complex near field within the Rayleigh distance and a far field thereafter. In the case of elliptical and rectangular apertures one dimension is usually much larger than the other, that is to say either the height of the aperture is appreciably greater than the width, e.g. for height finding radars; or the width is considerably greater than the height as for surveillance radars. The field can

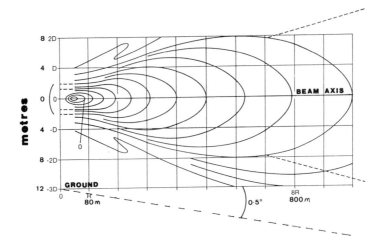

Figure 5.7 *Portraying beam elevation–ground level relationships on antenna field charts*

be described as having three identifiable regions as shown in Figure 5.3 where R_2 and R_1 relate to the Rayleigh distances R_2 and R_1 described below.

Figures 5.5 and 5.6 show the use of the letters A for the larger aperture dimension and B for the smaller aperture dimension as a reminder when using those diagrams. The field regions are:

1 Near field region up to the Rayleigh distance as explained earlier but with $R_2 = B^2/2\lambda$, where B is the smaller aperture dimension.
2 Far field region at a distance greater than $R_1 = A^2/2\lambda$ as for the circular aperture above where the power flux on axis is given by the far-field equation 1.
3 An intermediate field region between R_2 and R_1 where the power is spreading out in the 'B' direction but not in the 'A' direction.

In order to find the power density in the far field zone Figure 5.4 can be used, putting $R = A^2/(2\lambda)$. It is necessary to alter the vertical axis scaling of Figure 5.4 appropriately, as this is marked in terms of D the diameter of a circular antenna. This is done by putting $D = A$ for the case where the greatest dimension (A) is vertical e.g. a height finding radar and $D = A^2/B$ where the greatest dimension (A) is horizontal e.g. for a surveillance radar. The reference value L is:

$$L = P-T+G-10\log(4\pi R^2) \, dBWm^{-2} \tag{4}$$

Where the gain is not known, the following equation can be used. This corresponds to an efficiency of $\approx64\%$ for an elliptical aperture, or 50% for a rectangular aperture:

$$G = 8 + 10\log(AB/\lambda^2) \text{ dBi} \tag{5}$$

If the vertical aperture of a surveillance radar has been used for 'beam shaping', the above expression for gain may give an overestimate of several decibels so that some other method should be used.

For the intermediate field between R_2 and R_1 it is more difficult to estimate the power flux. In this region the power density on axis varies approximately inversely with distance $1/r$ since the power is spreading out in one direction but not in the other. The exact shape of the contours are complicated but the vertical sections of Figures 5.5 and 5.6 should give values within about 2 dB. The basis of these figures is that the power flux on axis varies as $1/r$ and for Figure 5.5 that the the radiation pattern in the vertical plane is fully formed. For Figure 5.6 it is assumed that the radiation pattern in the vertical plane is not formed at all.

Figure 5.5 is used where aperture dimension A is horizontal as for a surveillance radar and Figure 5.6 when the aperture dimension A is vertical as for a height finding radar. The reference power flux L_2 in dBWm^{-2} is given by:

$$L_2 = P - T + G - 10\log(4\pi R_1 R_2) \text{ dBWm}^{-2} \tag{6}$$

where $R_1 = A^2/(2\lambda)$ and $R_2 = B^2/(2\lambda)$

Example 2

Assume a surveillance radar (dimension A horizontal) with the following characteristics:

$P = 2 \text{ kW} = 33 \text{ dBW}; \quad f = 3000 \text{ MHz}; \quad \lambda = 0.1 \text{ m}; \quad G = 38 \text{ dBi}; \quad T = 0.48 \text{ dB};$
$A = 4.5 \text{ m and } B = 2 \text{ m}.$

Charts Figure 5.5 and Figure 5.4 are needed.

Step 1 Intermediate field
Calculate R_1 and R_2 and thence calculate L_2 as equation 6.

$R_1 = 101 \text{ m}$; $R_2 = 20 \text{ m and } L_2 = 26.5 \text{ dBWm}^{-2} \text{ or } 443 \text{ Wm}^{-2}$

Figure 5.5 should have the horizontal axis scaled in multiples of R_2 (20 m) and the vertical axis in increments of B (2 m). Note that the intermediate field as plotted is that up to the distance R_1 along the 'X' axis i.e. 101 m.

Step 2 Far field
$R = A^2/(2\lambda) = 101 \text{ m from step 1.}$

Using equation 4, and $R = 101 \text{ m}$ calculate $L = 19.5 \text{ dBWm}^{-2} \text{ or } 89 \text{ Wm}^{-2}$.

Scale figure 5.4 horizontal axis with multiples of R (101 m) and the vertical axis with $D = A^2/B = 10.1$ m (say 10 m).

Figures 5.8 and 5.9 show the results of the example 2 calculations, expressed in Wm^{-2}.

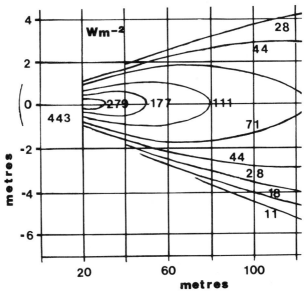

Figure 5.8   *Calculated values for the intermediate field in example 2*

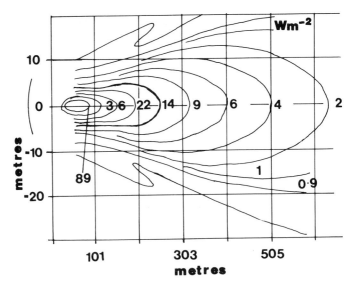

Figure 5.9   *Calculated values for the far field in example 2 (values below 1 Wm^{-2} not plotted)*

Use of a computer spreadsheet for calculations

The calculations used for the three normalised contour diagrams can easily be laid out in a computer spreadsheet so as to provide a quick method of producing results. By laying out parts of the relevant equations sequentially, the values of all the quantities of interest are readily available. These include the Rayleigh distance, wavelength, and the values of L in $dBWm^{-2}$ and Wm^{-2}. Figure 5.10 was produced from a well known spreadsheet and uses the equation for the far field of a circular aperture (dish) antenna.

The equation illustrated is equation 2 from earlier in this chapter:

$$L = P - T + G - 10\log(4\pi R^2) \ dBW^{-2}$$

The diagram shows the alpha and numeric cell references. All the calculations lie in column B B18, all other cells containing 'labels'. The numbers relate to example 1 earlier in this chapter and are left in cells B1, B3, B5, B6 and B7. This acts as a useful check on the calculation to detect any corruption and also to avoid the 'cannot divide by zero' message!

A new calculation just requires the relevant data to be inserted in the cells mentioned above and then B4 gives the Rayleigh distance, cell B11 the value of L (the reference level) in $dBWm^{-2}$ and cell B13 the value of L in Wm^{-2}.

The method can be used in the same way for the elliptical and rectangular aperture calculations by laying out the appropriate equations in the same way. In fact all the calculations in this chapter can be stored in spreadsheets and laid out for direct use.

A quick reference summary chart for the use of Figures 5.4 to 5.6 inclusive, is given in Table 5.3.

	A	B	C	D
1	freq MHz	3000.00	<--insert	
2	lambda(m)	0.10	------>	wavelength (m)
3	Diam	4.00	<--insert	diam (m)
4	R(metres)	80.00	------>	R=D*D/2*lambda
5	Power dB	30.00	<--insert	power
6	T loss dB	0.00	<--insert	attenuation
7	Gain dB	40.00	<--insert	gain
8		70.00	------>	P-T+G dB
9		80424.77	------>	4pi x RxR
10		49.05	------>	10*log of above
11		20.95	------>	=L dBW/sq m
12		2.09	------>	(antilog) L/10
13		124.34	------>	L=W/sq m

Figure 5.10   *Computer spreadsheet layout for antenna calculations*

Table 5.3   *Summary of the use of the calculation charts*

Aperture type	Field region	Diagram ref.	Calculation equation	'Y' axis scaling
Circular	Far	Figure 5.4	Equation (2)	No change
Elliptical/ rectangular	Far	Figure 5.4	Equation (4) Put R = $A^2/(2\lambda)$	Put D = A when A is Vertical; D = A^2/B where A is Horizontal
Elliptical/ rectangular	Intermediate ('A' dimension Horizontal e.g. as for surveillance radar)	Figure 5.5	Equation (6) $R_1 = A^2/2\lambda$ $R_2 = B^2/2\lambda$	No change
Elliptical/ rectangular	Intermediate ('A' dimension Vertical e.g. as for height finding radar)	Figure 5.6	Equation (6) $R_1 = A^2/2\lambda$ $R_2 = B^2/2\lambda$	No change

Useful approximations for microwave antennas

Approximations are useful in appropriate circumstances as cross-checks or in temporary circumstances until further data is obtained. They should not be a substitute for finding out the correct information, if it is available.

Where values such as L and L_2 occur below, the units applicable are as for the earlier reference in this chapter. Antenna gains given as a ratio and those given as dBi should not be confused. G_o is used to indicate ratios relative to an isotropic antenna and G for dBi. They are converted as follows:

G (dBi) = 10log G_o: Go = antilog G/10

Examples:

Ratio 1000; G = 10log 1000 = 30 dBi; G_o = antilog 30/10 = 1000

1 Circular apertures
(a) Gain of an 'ideal' antenna (excluding specialised types including satellite antennas):

D = antenna diameter (m); λ = wavelength (m) = 300/f (MHz)
G (dBi) = 10+20 log(D/λ)   (7)
(b) Gain of an antenna with an efficiency of $\simeq$64%:

$$G \text{ (dBi)} = 8 + 20 \log(D/\lambda) \tag{8}$$

Figure 5.11 shows the use of equation 8 for three circular aperture antennas.

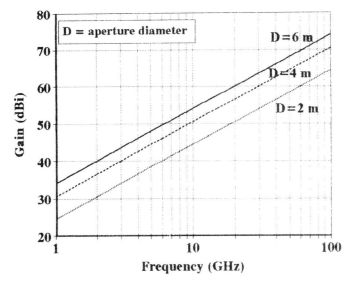

Figure 5.11 *Gain of circular apertures versus frequency for three aperture diameters*

(c) Maximum power anywhere on the axis where an antenna efficiency of ≈64% is assumed:

$S_{max} = +2.5$ dB or approximately 1.8 times the reference power flux L in equation 2.

Note that if the expression for gain for an antenna efficiency of 64% from (b) above is used in equation 2 to compute L, then:

$L = (2 \times P)/D^2$ which is a useful initial evaluation of L. Where the aperture area is used, the expression would be:

$$L = (1.57 \times P)/\text{Area} \tag{9}$$

and S_{max} (Wm^{-2}) would be $1.8 \times (1.57 \times P)/\text{Area} = 2.8 \, P/\text{Area}$ (10)

A conservative approach used by some organisations for the maximum power density anywhere on the axis in the near field is to use the above expression modified to assume an antenna efficiency of 100% giving:

$$S_{max} \text{ (Wm}^{-2}) = 4 \, P/\text{Area} \tag{11}$$

(d) Beamwidth estimate
Approximately $64 \lambda/D$ degrees for the 3 dB beamwidth. (12)

2 Rectangular and elliptical apertures

(a) Gain of an antenna for $\simeq 64\%$ efficiency (elliptical) or 50% (rectangular) apertures:

$$G \text{ (dBi)} = 8 + 10\log (AB/\lambda^2) \tag{13}$$

Where A and B are the aperture dimensions, as before. Figure 5.12 illustrates gain against the product AB for a range of frequencies, using equation 13.

(b) Maximum power anywhere on the axis where an antenna efficiency of $\simeq 64\%$ is assumed:

Using the expression for gain in equation 13 above in equation 6, the value of L_2 is now:

$$L_2 = P - T + 3 - 10 \log(AB) = (2 \times P)/AB. \tag{14}$$

Hence this value 2P/AB can give an initial estimate of L_2.

For the theoretical worst case (100% antenna efficiency) the maximum value anywhere on the axis is then 4P/AB. (15)

3 Miscellaneous

Occasionally it is necessary to make *ad hoc* calculations in respect of RF radiation incidents and these may involve open waveguides, horns etc.

(a) Approximate gain of open rectangular waveguides, pyramidal and sectoral horns with small aperture phase error:

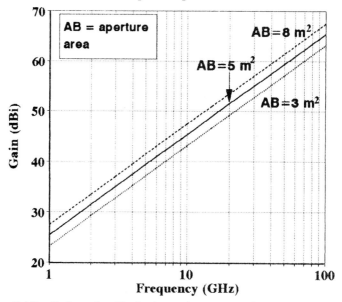

Figure 5.12 *Gain of elliptical and rectangular apertures versus frequency for three values of the product A × B*

$G_o \simeq 32 \, (AB)/\pi\lambda^2$ where A and B are the aperture dimensions in metres. (16)

(b) For circular waveguides:

$G_o \simeq 33 \, (r/\lambda)^2$ where r is the radius dimension (m). (17)

Other antenna calculations

Where it is really necessary to be concerned with the distribution of energy across the antenna aperture, there are a number of documents on the subject which can be used as sources of information and these include references 36, 69 and 70.

Simultaneous irradiation from several sources

In a situation where a person is irradiated simultaneously by a number of different sources, the INIRC recommendations[27] are as given below. Note the distinction between the cases of equations 19 and 20. This is due to the fact that below 10 MHz, SAR is of limited importance and induced current density produced in the human body is considered to correlate more directly with the biological effects resulting from human exposure.

Power density (at least one frequency above 10 MHz)

Let P_1, P_2 and P_3, P_n be the measured power flux densities for the relevant sources at the subject; let L_1, L_2, L_3, L_n be the corresponding permitted limits for the frequencies concerned:

Then $P_1/L_1 + P_2/L_2 + P_3/L_3 ... + P_n/L_n \leqslant 1$ (18)

Electric and magnetic fields (at least one frequency above 10 MHz)

Here E_1, E_2, etc., represent the measured field quantities (Vm^{-1} or Am^{-1}) and L_1, L_2, etc., represent the permitted limit for the frequencies concerned:

Then $(E_1/L_1)^2 + (E_2/L_2)^2 + (E_3/L_3)^2 ... + (E_n/L_n)^2 \leqslant 1$ (19)

Electric and magnetic fields (all frequencies below 10 MHz)

The letters and subscripts are exactly as for the previous equation.

$E_1/L_1 + E_2/L_2 + E_3/L_3 ... + E_n/L_n \leqslant 1$ (20)

Note: Other users of these formulae may apply different boundary conditions.

Near field measurements (>10 MHz to 30 MHz)

INIRC also identifies a formula to estimate the exposure rate in terms of the plane wave power density equivalents of the measured electric and magnetic fields over the frequency range 10 and 30 MHz (and in rare instances, up to 100 MHz). The rare instances are not defined. The calculation is based on the premise that the electric field component contributes about five sixths of the equivalent plane wave energy compared with the magnetic field component contribution of about one sixth.

The method used involves the measurement of the two field components, as required by most standards when making measurements in the near field and specifically at HF frequencies.

The equivalent plane wave power densities are then calculated for the electric and magnetic field values obtained and summed in the one and five sixths proportions referred to above. The result is then compared with the equivalent plane wave power density specified in the relevant standard.

Whilst the calculation is interesting, the author has not met many general cases where there would be any specific value in doing the calculation, other than cases such as tower climbing where a less conservative estimate is sought. Normally the field measurements would be evaluated separately against permitted values for those quantities in the standard.

The expression is:

$$5/6[(E^2)/377] + 1/6[(H^2) \times 377] \leq S_{eq}.(Wm^{-2}) \tag{21}$$

Where E is the measured electric field (Vm^{-1}); H is the measured magnetic field (Am^{-1}) and S_{eq} is the specified plane wave power density (Wm^{-2}). The expressions in square brackets will be recognised as those used to convert field quantities to equivalent plane wave power densities.

Peak power in a pulse

General

The need to limit human exposure to the peak pulse energy has been recognised for a long time but there has been a difficulty in establishing a mechanism for determining meaningful limits. The references to standards below can be followed up in Chapter 4. The relationships involved in pulse transmission were discussed in Chapter 1 and, in summary, are:

Duty factor DF = pulse duration t_p (secs) multiplied by the number of pulses per second n (Hz). This gives a number <1, a typical value being 0.001. It is easier to use the reciprocal which is 1000 in this example.

Mean power density × duty factor reciprocal = peak pulse power density:

$S_{mean} \times 1/DF = S_{pk}$

Thus for a permitted mean power density of $50\,Wm^{-2}$ and $1/DF = 1000$:

$S_{pk} = 50\,000\,Wm^{-2}$ $(50\,kWm^{-2})$.

Specifying peak pulse power density limits

The main methods used to specify a limit for S_{pk} are generally either to give specific peak pulse power density limits or to specify a limit for pulse energy (Jm^{-2}).

INIRC provides for peak pulse power density and the corresponding field components by means of a multiplier so that the normal mean power density limits in the standard are multiplied by 1000. For plane wave conditions, this corresponds to $\sqrt{1000} = 32$ times the electric and magnetic field components. This is the simplest possible method of tackling the problem.

The NRPB GS11 document specifies that for single pulses $< 50\,\mu s$, 'exposures in excess of $0.4\,Jm^{-2}$ shall neither be prolonged nor frequent'. This can be converted to peak pulse power density by dividing it by the duration of a pulse in seconds:

$S_{pk}\,(Wm^{-2}) = 0.4\,Jm^{-2}/t_p$

and in the previous example of $t_p = 20\,\mu s$ then:

$S_{pk} = 0.4/(2 \times 10^{-5}) = 20\,000\,Wm^{-2}$

A $2\,\mu s$ pulse would give $S_{pk} = 200\,000\,Wm^{-2}$.

Because there can be a wide variation in duty factor over different equipments and operating conditions, it is necessary not only to consider whether an exposure exceeds the permitted peak pulse power limitation but also to ensure that where the peak pulse power is acceptable, the mean power density is also acceptable.

Other methods of specifying peak pulse power density

The IEEE C95.1–1991 standard, provides expressions to calculate peak power density and also to limit the energy density where the pulse duration exceeds $100\,ms$, see Chapter 4.

Safety with moving microwave beams

The nature of moving beam systems

Microwave beams used for purposes such as radar surveillance, height finding and missile tracking may move in a constant or sporadic fashion. Conventional radar surveillance antennas normally rotate continuously at a fixed rate, typically 6 revolutions per minute, and constitute one particular case which can be examined.

Modern phased array antennas may be fixed with the beams being switched electronically or may rotate constantly as for a normal surveillance radar but have some form of electronic beam switching in elevation, either for the receive beam or for the transmit beam. If they have unpredictable transmit beam movements, they may need to be treated as other equipments with unpredictable movement patterns.

Radar height finders move round in azimuth and up and down in elevation. They usually park downwards and, if they are not arranged to switch off in this position, can produce a very hazardous area immediately around themselves. Their general movement is not predictable because of their function.

Missile and other tracking radars fall into the same category. In all such cases the problem is that the antennas could, during the course of proper operation, point at people and dwell there for an unspecified time.

There are, therefore' two basic cases which apply to both the mechanically moved and electronically moved beams and to combinations of the two:

1 Beams which rotate or scan an arc continuously at constant speed.
2 Beams which move without predictable patterns of movement.

The word 'predictable' here means that there is a pattern of movement which can be applied mathematically to the determination of the time-averaged mean power density exposure of a subject who is standing in the vicinity of the radar.

When a continuously rotating beam irradiates a person in its path, the duration of the exposure will only be a fraction of the total time taken for one revolution. Hence the mean power density will only be a fraction of that which would be measured at the location of the subject person, if the beam were stationary and aligned to irradiate that person.

In order to look at the methods used to calculate such irradiation, it is useful to consider two possible requirements. The most common one is to determine the safe boundary for people to work in the vicinity of the source.

The second requirement, which is just a variant of the first, is to determine whether people can work at a specific place or to investigate whether a person has been overexposed at a particular place.

The only difference between the two is the amount of work which might be involved as the setting of a general safety boundary around an antenna may involve measuring at a number of places whereas the second case is limited to a particular place.

Continuously moving beams

'Moving' here includes any form of scanning in a regular and repetitive way. The most common case is the antenna which rotates through 360°in azimuth continuously and at a constant rate. There are two cases to consider:

1 The exposed person is located at such a distance from the antenna that he is in the intermediate or far field of the antenna as Figure 5.13
2 The exposed person is in the near field of the antenna as Figure 5.14.

Intermediate or far field case

If we consider Figure 5.13 where the subject is located in the far field at point X, and assume that the beam rotates continuously at 6 rpm, it is obvious that X will be irradiated once per revolution. For the greater part of the revolution, there will not be any irradiation. It follows that the average power density over a single revolution seen at X is governed by a fraction which is related to beamwidth and the antenna rotational arc, here 360°. This factor is generally referred to as the 'rotational factor' which we can abbreviate to f_{rt}.

Since X is located in the intermediate or far field, the ratio involves the 3 dB bandwidth of the beam and the total angle of one revolution (360°). If the beam is now stopped and pointed directly at X, the stationary beam power density on the beam axis can be measured there. The product of f_{rt} and this measured power density is the average exposure of the subject at X and is usually referred to as the rotationally-averaged power density S_{rt}. An acceptable value of S_{rt} can be applied as a circular boundary round the antenna as shown in Figure 5.13.

Clearly this is the worst case. However in practice, allowance must be made for the fact that if there are a number of work places all at the same distance from the antenna, measurements made at some of these points may be very different if some places have metal masses which can cause enhancement of the power density by reflections.

This may have to be taken into account by making additional measurements with the beam stationary.

Example of a calculation at one place using angles:
Assume that the stationary measured value $S = 500\,\text{Wm}^{-2}$ at point X and that the beam 3 dB bandwidth is 3°. To find the rotationally-averaged power density the stationary value is multiplied by the rotational factor f_{rt} thus:

Rotationally-averaged power density $S_{rt}\,(\text{Wm}^{-2}) = f_{rt} \times 500$ and $f_{rt} = 3\,\text{dB}$ beamwidth (degrees)/rotation arc (degrees) = 3/360 = 0.0083

hence $S_{rt} = 0.0083 \times 500 = 4.15\,\text{Wm}^{-2}$ mean power density.

This must be compared with the permitted limit of the relevant standard for continuous exposure at the frequency concerned.

'Time' calculation method

It is possible to approach the same example from the time point of view.
Example of time calculation method:

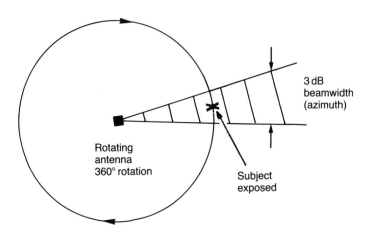

Figure 5.13 *Rotating beam power density time-averaging for a subject in the intermediate/far field*

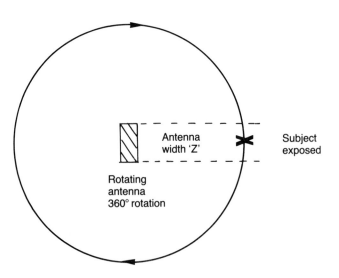

Figure 5.14 *Rotating beam power density time-averaging for a subject in the near field case*

Duration of one rotation at 6 rotations in 60 s = 10 s.

Duration of one exposure = (3°/360°) × 10 s.

$$= 0.083 \text{ s and } f_{rt} = \frac{\text{exposure time } (0.083 \text{ s})}{\text{rotation time } (10 \text{ s})} = 0.0083$$

Which is the same value for f_{rt} as in the previous calculation. It can be seen that the value of f_{rt} for one revolution suffices since using more than one rotation affects the numerator and denominator equally.

Note also that for a beam rotating for a full 360° or some smaller arc, the existence of sector blanking (see Chapter 8) does not affect the calculation as the ratio depends on the time duration of one sweep or rotation and not what the transmitter is doing during the sweep. The only difference in the general safety consideration is that there can, by definition, be no hazard in the blanked sector.

Note that a beam scanning less than 360° (say 180°) repetitively can still be dealt with by the two methods above since the first method uses the scan angle and the second the scan duration. The value of f_{rt} for 180° would be twice the value obtained in the 360° case, 3/180 instead of 3/360.

Alternatively, a repetitive 180° scan can be conceived as a 360° scan cycle in which the subject receives two exposures, one on each pass giving an f_{rt} of 3/360 + 3/360 = 3/180 as before.

For beams which move with different velocities at different angles relative to a datum but repeat the cycle exactly each sweep, the angle method is clearly unsatisfactory. It would be necessary to use the worst case and the time method. The worst case would be the exposure time at a position in the sector which corresponds to the lowest scan velocity (longest exposure).

Apart from the main beam, there will be sidelobes which need to be taken into account[72]. Usually these will be too low at these distances to be harmful but if they are thought or known to be significant they can be measured on a stationary basis or otherwise deduced as a percentage of the main beam measured power density on the basis of the specification of the antenna.

They could be treated as for the main beam in calculating subsidiary values of S_{rt} and summing all S_{rt} values to give the final calculation for the exposure of the subject. This is rarely necessary with high specification antenna systems.

In most cases where there is only some slight significance in the presence of sidelobes, it will be much easier to provide a comfortable contingency by doubling the 3 dB beamwidth in the initial consideration of f_{rt}, a practice used in most cases by the author as a general contingency factor.

The near field case

The previous examples dealt with exposures in the intermediate or far field of the beam. It is also necessary to consider the near field case. Figure 5.14 illustrates this case. The beamwidth is now determined as the largest dimension of the antenna aperture in the plane of rotation, shown as Z in the diagram.

$$\text{Now } f_{rt} = \frac{\text{aperture length in plane of rotation (Z)}}{2\pi \times \text{distance from antenna to the subject}}$$

Reference to Figure 5.14 shows that if the antenna is shown at the centre of a circle drawn such that X (the position of the exposed subject) lies on the circumference, then the distance of that person from the antenna is the radius of the circle. The denominator above is the circumference of the circle and $f_{rt} = Z/2\pi r$.

Example:
Assume that the measured power density at X in Figure 5.14 with the beam stationary is $650\,\text{Wm}^{-2}$ and that rotation is 360°. Point X is 20 m from the antenna and the aperture length (Z) is 5 m.
$f_{rt} = 5/(2\pi \times 20) = 0.0398$
and $S_{rt} = 650 \times 0.0398 = 25.9\,\text{Wm}^{-2}$

Where the arc swept is less than 360°, the denominator should be reduced proportionately on the basis of actual arc°/360°. Again, no notice should be taken of the fact that part of the arc is sector-blanked as it is the arc sweep time that is the determinant and not the nature of the activity of the transmitter.

Example:
Assume that the scan is 180° and the other data as the previous example:

$f_{rt} = 5/(2\pi \times 20 \times 180°/360°) = 0.0795$
and $S_{rt} = 650 \times 0.0795 = 51.7\,\text{Wm}^{-2}$

This near field method of calculation is conservative, as it should be from the safety point of view, overestimating the value of S_{rt}.

Practical examples

The more general case is likely to be to determine the stationary beam power density required to give a permissible rotational power density S_{rt}. This would enable a hazardous area limit to be applied to the equipment, for example, in the form of a circular perimeter limiting access. It is important, as noted previously, that any metal masses swept by the beam which might provide fields higher than those calculated at the distance in question, should be investigated.

Example 1: Outside the near field

It is proposed to locate personnel outside the near field of a continuously rotating radar with a uniform rotational velocity. The rotation speed is 6 rpm and the 3 dB beamwidth is 4°. What stationary beam power density will correspond to a permitted exposure level of 32 Wm^{-2}?

S_{rt} = 32 Wm^{-2}; let S_0 represent the required stationary beam mean power density.

and $S_0 = S_{rt}/f_{rt}$

The beamwidth is 4° so that the rotational factor is:

f_{rt} = 4°/360° = 0.0111

Now S_0 = 32/0.0111

S_0 = 2883 Wm^{-2}

It would now be necessary to measure the stationary beam at the relevant height above ground e.g. 0 to 2.5 m for personnel clearance at ground level, to establish the distance at which the power density is not greater than 2883 Wm^{-2} or refer to measurements already made at various distances from the radar.

Note that as the stationary power density is 2883 Wm^{-2}, many times the permitted levels at microwave frequencies in any of the standards mentioned in Chapter 4, It will be clear that the radar could not be allowed to radiate when stationary e.g. if rotation failed, since it may point at people and expose them to this excessive level. It would also be hazardous for the surveyor – see Chapter 9!

Example 2: In the near field

A task is to be undertaken at 10 m from a continuously rotating antenna in the near field. The rotation rate is 6 rpm. The permitted limit for exposure at this frequency is 40 Wm^{-2}. The antenna aperture length in the plane of rotation is 5 m. The measured power density at the proposed place of work with the beam stationary is 1200 Wm^{-2}.

1 Would it be permissible to work in this position?
2 If not, what distance would be permissible?
3 Alternatively, what transmitter power reduction would be required to enable the work to be done?

In the near field, the beamwidth is assumed to approximate to the antenna aperture length in the plane of the radiation.

$$f_{rt} = \frac{\text{aperture length in the plane of rotation}}{2\pi \times \text{distance from antenna to exposed person}}$$

Consider the subject as being on the circumference of a circle whose radius is the distance from the antenna to the exposed person, the antenna being by definition at the centre of the circle. The denominator above is then the circumference of the circle.

Now $f_{rt} = 5/(2\pi \times 10) = 0.0795$

and $f_{rt} \times 1200 = 95.4\,\mathrm{Wm}^{-2}$.

1 This exceeds the specified $40\,\mathrm{Wm}^{-2}$ and it would not be permissible to work at this place without some other method of reducing the exposure.
2 The value of the power density in the stationary beam at the workplace needs to be reduced. Let the new value = S.

   $f_{rt} \times S = 40$; re-arranging, $S = 40/0.0795$

   Hence $S = 503\,\mathrm{Wm}^{-2}$. The new distance would have to be established by measurement.
3 Since the previous measured stationary beam power density at the same place was $1200\,\mathrm{Wm}^{-2}$, and the required limit from (2) above is $503\,\mathrm{Wm}^{-2}$, the required power reduction ratio will be: $503/1200 = 0.42$ or $-3.77\,\mathrm{dB}$.

Note that the figures used in the above example are chosen to illustrate calculation methods. Caution is needed in authorising such exposures in the near field and it would be important to ensure that the fields have been adequately explored.

Special cases

Some surveillance radars use a single rotating platform with two antennas mounted back-to-back so that a person standing at a point will be irradiated twice in each revolution, once by each radar. The frequencies will usually be different and the permitted exposure limits for the two cases may also differ. It will be necessary to perform calculations for $S_{rt}(f_1)$ and $S_{rt}(f_2)$, where f_1 and f_2 identify the power densities for each of the sources.

If the permitted limits for the two frequencies are the same, the two S_{rt} values can be added. The total should not exceed that lant. Otherwise they are treated as indicated earlier under the heading 'multiple irradiations' on the basis of the sum of the partial fractions.

General limitations

In the foregoing text there are some important aspects to be considered and these are outlined below.

Where the elevation of a rotating or scanning beam is fixed but can be changed, it is important that the setting for the measurement of the stationary power density required in connection with the above calculation methods should be the worst case from the point of view of the potential exposure of people and should include consideration of the nature of the potential work i.e. at ground level, on a mount, tower or gantry.

If clearance is only required for work at ground level then it should be endorsed as such e.g. 'personnel clearance to 2.5 m height from ground only'. Clearances for higher working need to be definitive and specific e.g. 'clearance to work on the lower platform only of gantry no. 6'.

Calculations of the type described above cannot take into account the non-uniformity of the environment around a rotating or scanning antenna such as might arise if there are potential sources of power density enhancement at particular places. These may involve metal masses, potentially resonant objects including discarded antennas, parts of new antenna structures being assembled and the like.

Here it may be necessary to treat each proposed workplace, even though it may be the same distance from the antenna as in a previous assessment, as a new survey task.

Alternatively, a large contingency, at least 6 dB, can be added to the previous calculation to cover other locations at the same distance from the antenna, and some measurements made at the location of possible enhancements to confirm the adequacy of the contingency.

Whilst the calculations outlined above relate to one source of radiation, in the real world there may be situations where the subject is simultaneously irradiated from other sources, scanning or fixed and it will, therefore, be necessary to establish these exposures individually and sum them by the methods given earlier in this chapter.

Irregularly moving beams

There is a straightforward 'rule' which emerges from consideration of the possible events which can occur with beams which have unpredictable modes of operation as far as the potential irradiation of people is concerned.

If a person can be subject to exposure to a power density which exceeds that specified in the relevant standard over the specified time averaging period (six minutes normally) then this is, by definition, unacceptable.

If this can occur by a beam dwelling in such a position as to continuously irradiate a person either in normal operation or due to a beam scanning failure then the situation is deemed unacceptable. Practical cases may include tracking radars, height finders and even constantly rotating antennas if the rotation can fail without shutting the transmitter off. There are other possible cases with some types of phased array antennas.

In such cases, the logical conclusion is that such beams should be assessed on a stationary basis. It is worth noting that some organisations also apply an additional empirical restriction so that if the individual is exposed for a time exceeding 1 second as the antenna scans, the beam is assessed as a stationary beam, i.e., no reduction allowed for rotation, even though the rotation is at a constant rate.

HF, UHF and VHF antennas

Antennas for these frequencies include whip and rod types, dipoles and wire arrays. The type of antenna will depend on the nature of the system, fixed HF transmitters often having vast arrays of wire antennas occupying much space. Mobile and some shipborne HF systems where space is at a premium or where mobility is important often have rod or whip types. HF arrays do not lend themselves to simple calculations of the type discussed in the previous paragraphs for microwave beams but some rough estimates can be done for whips, rods dipoles and similar 'compact' antennas.

VHF and UHF equipment uses antennas ranging from simple whip types for mobile work to dipole-based arrays with directors and reflectors such as the domestic television antennas, Yagi and log periodic arrays. For VHF and UHF broadcasting, circular modular arrays fitting round towers are often used.

For these lower frequencies the field zones that are recognised are:

1 The non-radiating reactive near field close to the antenna.
2 The near field radiating zone.
3 The far field radiating zone.

It should be noted that the reactive near field (induction field) becomes more significant as the frequency is reduced. Whilst this field has the non-radiating connotation, it involves RF energy just as the radiating fields do, and that energy can harm people in the same way as radiated energy.

The last two (near field and far field) are both radiation fields and there are a number of formulae for the calculation of these distances:

Boundary between the induction field and the radiation field (induction field < 10%)

$$d = \lambda/1.6\pi \text{ or approx. } \lambda/5 \tag{22}$$

Near field/far field relationship

$$d = \lambda/2\pi \text{ or approx. } \lambda/6.28 \tag{23}$$

d = distance in metres in both cases when λ is in metres.

Note that in the literature generally, the near field/far field division is often just referred to as 'approximately $\lambda/2$'.

Field intensity from antennas in free space (in the direction of maximum radiation)

$$S = \frac{PG_o}{4\pi d^2} \text{ (Wm}^{-2}) \tag{24}$$

Where
P   = transmitter output power (W)

G_o = gain ratio (for isotropic radiator G = 1)
d   = distance from antenna in metres

The electric field strength will be:

$$E = \frac{\sqrt{(30PG_o)}}{d} \; Vm^{-1}$$

and the magnetic field strength will be:

$$A = \frac{\sqrt{(PG_o)}}{68.83d} \; Am^{-1}$$

It can be useful to make a table of values for equation (24) for general use. Table 5.4 is an example for power density from a vertical antenna, calculated for a number of distances and different values of gain. Similar tables could be produced for the electric and magnetic field strengths. Such tables can be run out for different powers, and gains, using the 'make table' facility of a computer spreadsheet.

Table 5.4   *Power density (Wm^{-2}) versus distance from vertical antenna for gains of 0 to 3.5 dBi*
Power = 500 W

Gain (dBi) ⇒	0	1	2	2.15*	3	3.5
Gain ratio ⇒	1.0	1.26	1.58	1.64*	2.0	2.24

Distance (m) ⇓						
0.5	159.2	200.5	251.5	261	318.3	356.5
1	39.8	50.1	62.9	62.3	79.6	89.1
1.5	17.7	22.3	27.9	29	35.4	39.6
2	9.9	12.5	15.7	16.3	19.9	22.3
2.5	6.4	8	10.1	10.4	12.7	14.3
3	4.4	5.6	7	7.3	8.8	9.9
4	2.5	3.1	3.9	4.1	5	5.6
5	1.6	2	2.5	2.6	3.2	3.6
6	1.1	1.4	1.7	1.8	2.2	2.5
7	0.8	1	1.3	1.3	1.6	1.8
8	0.6	0.8	1	1.1	1.2	1.4
9	0.5	0.6	0.8	0.8	1	1.1
10	0.4	0.5	0.6	0.6	0.8	0.9

*Dipole

Safety calculations for structures involving flammable vapours

Introduction

The potential hazard involved where flammable vapours are used or stored in places where significant RF fields are present is fairly well known.

The real problem is to determine what field levels are significant in a given situation. The mechanism involved is that conductors, which might be pipes, girders, wires and other structural elements, collect energy from the field. Where there is a suitable discontinuity such that a spark occurs, any flammable vapours present may be fired.

The elements required are:

1 A sufficiently strong RF field.
2 Conductive structures capable of having potentials induced in them.
3 A discontinuity in the conductive elements capable of providing a 'spark gap'. This must be located in a place where flammable vapours are present.

Conductive structures are not difficult to visualise and, apart from basic structural elements, include metal pipes, tanks, metal supports and similar items. Installations used for industrial processing and storage of petroleum and other substances having flammable vapours, flammable gases such as natural gas and the many forms of industrial gases used in everyday life fall into this category.

The basic mechanism is that the conductive elements act as an antenna and collect RF energy, the magnitude of which will depend on the field and the effectiveness of the structure as an antenna.

British Standard BS 6656 1991 [31] is the latest standard in the UK to cover this subject and is used here to illustrate the basic calculations involved. No international standard on this subject is known to the author and many standards or similar documents tend to be military ones.

Basic approach

The basic approach of the standard is to detail a flow chart sequence of steps to ensure consideration of all the relevant aspects. This is not covered here, but the basic calculations involved are discussed in the following paragraphs. The standard covers frequencies from 15 kHz to 35 GHz. The pick-up characteristics of structures may result from parts of the structure acting as a monopole or from loops formed by the structure.

At frequencies below 30 MHz, to evaluate the power which can be extracted, structures are treated in terms of the effective loop formed by the conductive members as loops are considered to be the more efficient receiving antennas at these frequencies. For example, a pipe or series of pipes may constitute a loop and the 'loop length' can be measured and used in calculations.

Probably the simplest loop to conceive is a crane with a conductive connection (chain or wire) to the lifting hook. By definition the crane has a discontinuity when the hook is not touching anything (see Figure 5.15).

Because a crane is a particularly effective antenna, the standard treats this as a special case. Other conceptually simple loops include road tankers with

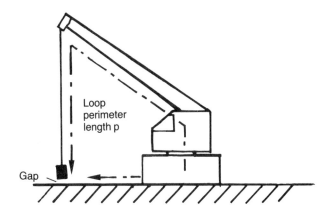

Figure 5.15 *Crane 'pick-up loop' illustration*

an overhead metal discharge arm and metal walkways across the top of two or more metal storage tanks, where the loop is tank-walkway-tank.

Sometimes a loop may turn out to be resonant (about one half-wavelength) or may become resonant when alterations are done to the conductive structure. In some cases it may be possible to detune the loop with added reactance, so reducing the effective pick-up.

At frequencies of 30 MHz and above, the maximum extractable power is treated on the basis of a half wave dipole with an added contingency of 10 dB to cover field enhancement, antenna gain and other effects for frequencies above 200 MHz.

Maximum extractable power-below 30 MHz

$$\text{If} \quad p/\lambda < 0.4: P_{max} = 702 \left(\frac{E^2}{f^2} \right) \left(\frac{\lambda}{p} \right)^{3.5} \tag{25}$$

$$\text{and} \quad p/\lambda > 0.4: P_{max} = 28.4 \left(\frac{E^2}{f^2} \right) \tag{26}$$

Where:
E = effective field strength (Vm^{-1})   f = frequency (MHz)
λ = wavelength (λ = 300/f) P_{max} = max. extractable
                                            power (Watts)
p = internal perimeter of the loop type structure in metres

Example:
Assume: f = 15 MHz   λ = 20 m   E = 30 Vm^{-1}
p = 5 m; hence p/λ = 5/20 = 0.25 and the first expression is the appropriate one:
P_{max} = 702 E^2/f^2 (p/λ)$^{3.5}$ = 702 × (900/225) × (0.25)$^{3.5}$

P_{max} = 21.9 W

If we keep all values as above except for p which we increase to 18 metres so that p/λ is 18/20 = 0.9 which is greater than 0.4, then the second expression above gives:

P_{max} = 28.4 × (E^2/f^2) = 28.4 × (900/225)

P_{max} = 113.6 W.

Maximum extractable power – 30 MHz and above

The expression used here is P_{max} = 311E^2/(f^2 + 9000)   (27)

And the meanings of the symbols used is as before. This expression is easy to manipulate, especially when concerned with only one frequency, since it can then be rewritten as:

P_{max} = kE^2 where k = 311/(f^2 + 9000)   (28)

Which is a constant for the frequency concerned.

Example:
If only one frequency is involved, say 50 MHz, k = 0.027 and P_{max} = 0.027E^2. This could be plotted for a number of values of E if required.

Thus when   E = 50 Vm^{-1}, P_{max} = 67.5 W
            E = 20 Vm^{-1}, P_{max} = 10.8 W.

We now have a means of calculating the extractable power from a defined situation. It should be noted that we are concerned with peak power for pulsed transmissions, peak envelope power for SSB and otherwise carrier power so that the electric field strength used should reflect this.

The next step is to see what consequences arise from this extractable power.

Characteristics of flammable vapours

The standard refers to a classification system for flammable vapours which groups them according to the degree of hazard.

For the purpose of providing examples, the three groups referred to are as shown in Table 5.5 which also shows a representative gas for each category. There is an extensive list of substances in the standard, the tabulations being obtained from BS 5501 pt1 1977[71] a translation of the EEC European standard EN 50014.

Table 5.6 shows the RF power thresholds for gas groups 1 and 2A, 2B and 2C in terms of watts with an associated averaging time. Extracted power

Table 5.5 *Gas groups and representative gases
(Courtesy of the British Standards Institute, London)*

Gas group	Representative gas
1 and 2A	Methane
2B	Ethylene
2C	Hydrogen

Table 5.6 *Gas groups and threshold powers
(Courtesy of the British Standards Institute, London)*

Gas group	Structures other than cranes	Cranes only
	Threshold power and averaging period	Threshold power and averaging period
1 and 2A	8 W averaged over 100 µs thermal averaging time	6 W averaged over 100 µs thermal averaging time
2B	4 W averaged over 100 µs thermal averaging time	3.5 W averaged over 100 µs thermal averaging time
2C	2 W averaged over 20 µs thermal averaging time	2 W averaged over 100 µs thermal averaging time
Note	Source impedance assumed = 3000 Ω	Source impedance assumed = 7500 Ω

from the previous paragraphs is now compared with these limits for the relevant gas group. Note the different figures for cranes. The examples below do not include cranes.

If the maximum extracted power exceeds these limits, it is then necessary to look to a reduction of the value of E (the electric field) in the previous calculations in order to stay within the limits. If we take the last example above, where for a field of $20\,Vm^{-1}$, $P_{max} = 10.8\,W$. This would not be satisfactory for any of the categories in Table 5.6 and must be reduced.

We can do this easily by putting the original expression equal to the limit with which we are concerned, say the group 1 and 2A limit of 8 W. It will be remembered that we had a simplified expression for a single frequency:

$P_{max} = 0.027\,E^2$

So if we put $P_{max} = 8\,W$, we have $0.027 \times E^2 = 8\,W$ and by solving for E we can meet the table limit.

$E^2 = 8/0.027$ and $E = 17.2\,Vm^{-1}$.

Most people would probably aim to be 10% or 20% below the tabulated limits to allow for the uncertainty of the field measurement and would, consequently, use perhaps 6 W rather than the 8 W above, giving about $15\,Vm^{-1}$.

Pulse transmissions

Where pulse transmissions are used and the pulses are short relative to the thermal ignition times in Table 5.6 (less than half of these times) and the interval between pulses exceeds the thermal ignition time, it is recommended that the energy in a single pulse is a better criterion.

Table 5.7 lists threshold pulse energies for the same gas groups as Table 5.6. If we take the example from the previous paragraph where P_{max} was 8 W

Table 5.7 *Gas groups and threshold energies*
(Courtesy of the British Standards Institute, London)

Gas group	All structures – RF energy thresholds (μJ)
1 and 2A	7000
2B	1000
2C	200

for an electric field strength of $17.8\,Vm^{-1}$ at 50 MHz, and suppose this to be a pulsed transmission meeting the criteria just given, the figure of $17.8\,Vm^{-1}$ is now, of course, to be derived from the peak pulse power and not the mean power used for CW.

If Z_{max} is the maximum extractable energy from one pulse in microjoules and t_p is the pulse duration in microseconds, we have:

$$Z_{max} = P_{max} t_p$$

Assuming a value of $t_p = 3 \, \mu s$, $Z_{max} = 8 \, W \times 3 \, \mu s$

Hence $Z_{max} = 24$ microjoules, which would meet the requirements of Table 5.7. Of course the value of $17.8 \, Vm^{-1}$, is a very low one for a pulsed transmission peak pulse power, corresponding to a plane wave peak pulse power density of $0.84 \, Wm^{-2}$. A more realistic approach is to make Z_{max} equal to one of the Table 5.7 values, say, $7000 \, \mu J$, and find the limiting value of P_{max}.

Thus $P_{max} = Z_{max}/t_p = 7000 \, \mu J/3 \, \mu s$

Table 5.8 *Examples of hazard radii associated with specific systems (Courtesy of the British Standards Institute, London)*

Type of transmission	Frequency	Power watts	Mod. type	Factor (see note 1)	Antenna gain dBi/ polarisation	Radii of vulnerable zones in metres		
						Gas groups 1/2A	2B	2C
Broadcast	198 kHz	500 kW	AM	2	4 V	4300	5500	7200
Tropospheric scatter	900 MHz	10 kW	FM	1	46 V/H	900	1200	1500
Radar (military)	1 to 3 GHz	6 MW peak pulse	P	1	45 V/H	500	1400	3000
Radar (civil)	1 to 3 GHz	2.5 MW peak pulse	P	1	39 V/H	330	860	2000
VHF/UHF base station	98 to 400 MHz	125 W	FM/ AM	4	5 V	20	25	30
HF land communications (note 2)	4 to 25 MHz	30 kW	SSB	4	18 H	630	720	830
					18	260	290	330
HF communications (note 2)	1.6 to 25 MHz	30 kW	SSB/ FSK[12]	4	12	8700	10900	13700
						190	230	260

Notes:
1 This factor is to allow for multiple sources; the transmitter power is multiplied by it.
2 Where two types of modulation are shown, the worst case is used. The two sets of distances relate to the ends of the frequency band.
3 P = pulse transmission; for other modulation terms see Appendix 1.

Hence $P_{max} = 2333\,W$ and the corresponding electric field value from equation 28 would be $294\,Vm^{-1}$, a possibly more realistic figure. The general rationale is:

1 Find the relevant gas group and limit threshold energy (Table 5.7). Derate further if required for risk reduction
2 Compute P_{max} = chosen Z_{max}/t_p
3 Using the relevant formula for P_{max} from previous paragraphs, calculate E, the maximum electric field value:
   For example, using equation 27:

$$P_{max} = 311E^2/(f^2 + 9000) \text{ and } E^2 = P_{max}\,(f^2+9000)/311$$

Other aspects

The standard also gives formulae for the calculation of field strengths from transmitters. In the situation where measurements are made, these calculations may be of little interest but they might be of value when considering a new transmitter site before access for measurement is obtained. Table 5.8 gives a few examples from a large table of examples of the radii of vulnerable zones.

The table is only included here to give a general impression of the range of hazard distances typical of the systems portrayed. For practical use it is important to see the qualifications applying to the table which appear in the standard. The standard contains a great deal of other useful tabulated data.

A companion standard (BS6657)[32] deals with the corresponding problems of the firing of commercial electro-explosive devices e.g. commercial detonators, by RF radiation. Most non-commercial explosives are military and are covered by specific military documentation which is likely to be classified information.

6
RF radiation measuring instruments

Purpose and classification

The primary purpose of safety measurements is to verify compliance with safety guides and standards. The type of measurements made will therefore be related to the parameters specified in such documents. Some RF test equipment purchasers have their own specifications which include safety tests specific to their own requirements. This may, on occasions, require additional measuring equipment.

Groups of measuring equipment can be identified according to their application, though these groups are not mutually exclusive.

Portable measuring instruments for field use

Instruments are needed to carry out surveys indoors and outdoors, involving the measurement of power density, electric field and magnetic field quantities. Additionally, new techniques and instruments such as limb and contact current measuring meters are becoming available. The requirements for instruments will include ergonomic factors such as ease of carrying and manipulating the instrument and low weight to reduce wrist strain when using the equipment for long periods.

In many cases use may include climbing towers and structures, operating the equipment on the way up and down. Here, minimisation of the use of the hands is important and can often be achieved by a chest mounted or side slung carrying case for the instrument meter unit, so that only the probe has to be held in the hand.

There will usually be a tendency to use the same instruments for the inside and the outside work to avoid excessive investment in equipment unless there are good reasons for choosing different equipment. Surveys may include research and development measurements on new designs and the investigation of complaints about suspected irradiation.

Systems for laboratory and production use

There are also some systems which are not easily moved about in the sense mentioned above due to the weight or bulk involved and are more suitable for fixed use in a test bay or in a safety check-out system. These, typically, provide central control of a number of sensors and may, increasingly, tend to offer some degree of automation of the testing of volume products such as microwave ovens.

Personnel safety monitors

These include equipment designed either to be worn as a personal monitor or fixed in a room or enclosure and intended to warn people in the room of fields which exceed a preset threshold.

Such monitors are often used in equipment maintenance activities such as ground and aircraft transmitter servicing bays and in the comparable military ground, ship and avionics activities with radio, radar and jamming transmitters.

Some types of fixed monitors lend themselves to some degree of safety control by providing external electrical signals or a contact closure when the preset alarm level is reached. This can be linked to warning alarms, illuminated warning signs or even physical barriers such as an electrically-locked gate.

There is a considerable development activity in this field at the moment and the number of equipments falling under this heading is rapidly increasing.

Induced current measurement systems

These are equipments designed to measure currents induced in the human body, for example in the limbs, during exposure to fields, and contact current instruments intended to measure the currents that can be experienced when touching metal masses located in an RF field. The standards governing these subjects are discussed in Chapter 4.

Ancillary equipment

Occasionally, receivers and spectrum analysers may be needed for diagnostic work such as the identification of interfering signals giving an out-of-band response during field measurements. Equipment in this category is well-established general purpose test equipment which needs no detailed discussion here.

Other ancillary equipment includes small test sources intended to test that RF survey instruments function correctly. This should not be confused with checking calibration, for which the test sources would be inappropriate. These sources are very useful since the most likely cause of a failure with most instruments tends to be the probe cable or a discharged battery. They are commercially available.

It should be noted that it is often necessary to ensure that test sources do not themselves create local electro-magnetic interference to operational systems at the place of use. In such cases, the sources are often fabricated with a half-spherical or conical aperture into which the probe is located when using the test source.

Nature of instruments for RF radiation measurements

Construction and operation

There are now a number of established commercial suppliers of specialised RF radiation measuring instruments for RF safety measurement work covering power density, electric field strength and magnetic field strength.

The suppliers are, for the most part, located in the USA but usually have agents in other countries.

Some of the products of these companies are illustrated later in this chapter. Over the last twenty years or so there has been a steady development of such instruments and a gradual extension of the frequency range of the wideband instruments. This still lags transmitter development as higher frequencies come into use, and in these cases there is a problem in carrying out safety measurements. It should perhaps be noted that some suppliers of modern wideband instruments can provide calibrations for frequencies higher than the specified upper frequency, so that it may be worth discussing the matter with them.

The basis of such instruments is shown in a much simplified theoretical form in Figure 6.1 and involves a receiving antenna, a square law detector to produce a true r.m.s. output, a suitable amplifier and an indicator. The antenna and detector is often referred to, in the generality, as a sensor. The indicator is usually a conventional moving coil meter. Often there is an output available to feed external equipment either directly, or by accessories such as a fibre optics link, to avoid the problem of pick-up on extended wires.

The detector used is either a thermocouple or a diode circuit. The general physical realisation is normally that of a probe sensor connected by cable to a meter unit. The meter unit has batteries, a meter, switching facilities and circuits to process the probe signal. More recently designed instruments may have additional facilities and these are described later.

A schematic diagram tends to conceal the technical problems involved in the design of the instrument portrayed. The key aspects of the design are:

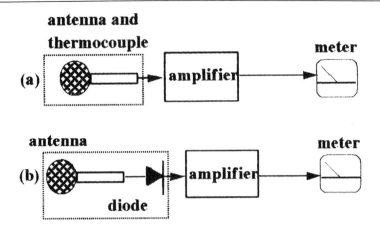

Figure 6.1 *Schematic illustration of instrument concepts*

1 Design of the antenna, the effective bandwidth covered and the choice of type of detector – diodes or thermocouples.

2 RF isolation of the antenna from the rest of the instrument so that, in the theoretically ideal situation, the sensor (antenna and detector) appear as totally isolated and of infinitely small effective size so as to ensure very loose coupling with the field. The objective is to minimise perturbation of the field and errors due to gradients across the probe in situations where there are steep spatial gradients in the field.

3 Protection from interference due to frequencies outside the measuring bandwidth of the instrument, i.e., 'out of band responses' giving rise to readings which may be thought to result from the in-band (wanted) field under test.

4 The field which is being measured should not be able to induce currents in the instrument other than via the sensor. In particular, it should not be able to induce currents in the lines connecting the sensor to the meter unit. Nor should it interfere with any circuitry in the meter unit. The EMC aspect is becoming more important as extra digital circuitry is built into the meter unit to provide new facilities.

Sensor antennas

The antenna is normally in probe form, that is to say that it is sited on the end of a tube which constitutes the handle. The antenna and detector arrangement is a very critical part of the design.

The requirements are very exacting since the flatness of the frequency response over the bandwidth involved (typically 0.3 GHz to 40 GHz on the wideband types) determines, in a large part, the measurement specification for the instrument.

The antenna will be arranged to measure either the electric or the magnetic fields at lower frequencies but in the wideband instruments covering the microwave region, the electric field is used exclusively above 300 MHz. It is important that an electric field sensor should not be affected by the magnetic field and vice versa. Lossy elements are used for the wideband antenna systems, using thin film techniques.

The basic antenna types for electric and magnetic field measurements are illustrated in Figure 6.2. This shows a single electric field antenna element in

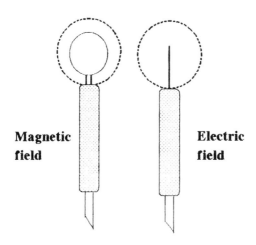

Figure 6.2 *Sensor non-isotropic antennas*

a typical probe configuration and alongside it a single magnetic field pick-up loop. The single pick-up element will only respond to the polarisation corresponding to that of the source, for example, when the probe electric field element is parallel to the source electric field.

Such antennas are known as non-isotropic, that is to say they do not respond to all the energy in a field containing signals with more than one polarisation, such as that which involves reflections from local conductive objects including the ground, or to the elliptical polarisation in the near field. As a result, they will not give an indication of the total field present.

The most frequently used systems are isotropic and consist of three orthogonal elements with the outputs joined together to give a combined output which can be shown to correspond to the average value of the sum of all the impinging RF energy components. They cannot, of course, provide phase information.

Isotropic sensor elements are illustrated in Figures 6.3 and 6.4 in the form of models in order to increase the clarity of presentation. Figure 6.3

Figure 6.3   *Sensor isotropic antennas – electric field*

Figure 6.4   *Sensor isotropic antennas – magnetic field*

illustrates an isotropic electric field sensor consisting of three lossy antenna elements and Figure 6.4 depicts an isotropic magnetic field sensor comprising three loop antennas. The method of joining together the three electric field elements may be end connection as illustrated, whereas for dipole elements they may be centre connected.

Practical instruments for magnetic field measurement often use square loops instead of the circular ones modelled above. It should be noted that the loop arrangement where the three loops have a common geometric centre is very important to avoid one loop shadowing another.

This can be best understood by considering an alternative physical realisation of the loop system where the loops are mounted on three mutually adjacent faces of a cube.

It can then be seen that as the probe is moved about, there can be shadowing of one loop by another, affecting the instrument indication and causing a departure from the isotropic state.

Sensor detectors

In instruments of the thermocouple-based types, the thermocouple elements are, typically, thin film depositions of antimony and bismuth on a thin plastic substrate within the probe antenna structure as illustrated in Figure 6.5 for an electric field probe element from one manufacturer. In this

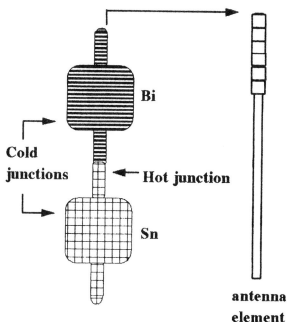

Figure 6.5 *Typical thermocouple elements (Courtesy Loral Microwave-Narda Company)*

example, these fold over into a stack on the electric field antenna. By this means the small spacing between the cold and hot junctions minimises thermal drift and interconnection is facilitated.

For the magnetic field probe, the thermocouples are fitted in the individual loops. For both types of probe the amplifier is located close to the antenna, usually being fitted in the probe handle.

The connecting leads are high resistance to provide a degree of isolation of the sensor and a loose coupling into the field. The amplifier conditions the signals passed to the instrument unit by reducing the impedance and thus reducing effects caused by probe cable flexing. Similarly, in diode-based instruments, the diodes are located in the probe antenna structure.

The thermocouple array will give an output corresponding to true r.m.s. power regardless of wave shape. It is most useable above about 300 MHz and is commonly used in wideband probes. At the lower frequencies it is less efficient and diodes may be used.

The diode will give a square-law output over a defined range of operation but can give rise to errors of appreciable magnitude with amplitude modulated waves, pulse transmission or with two or more equal amplitude signals present simultaneously. This is due to the diode operation moving from the square law into the linear detection region. For two equal signals present simultaneously, the theoretical error is 3 dB based on the squaring of the sum of the signals rather than summing their squares.

With short duration low duty factor pulse modulation such as used in radar equipment, considerable error can result, perhaps up to 30 dB. However most manufacturers of instruments using diode detectors have patented circuits which, it is claimed, can substantially reduce this error.

One manufacturer discusses in an instrument handbook the case of two or three equal signals and quotes tests carried out on the instrument concerned with two and with three equal signals present simultaneously.

The tests were carried out by an independent laboratory and showed the maximum error as 1.76 dB. Another manufacturer claims a maximum error of 1 dB even with low duty factor short pulse transmission. It should be noted that not all probes are designed for short pulse work of the radar kind.

Both diodes and thermocouples can easily be destroyed by excessive overloads, that is to say, in excess of the maximum values specified by the manufacturer. Diodes will generally stand more overload than thermocouples but the latter have improved in this respect with developments in fabrication techniques.

Variations in constructional methods

Not all instruments use probes with flexible leads. Some use a probe rigidly fixed to a meter unit thus reducing connecting lead length but reducing flexibility in handling. This is not always a disadvantage, for example, if the instrument has a static utilisation in a test rig.

At least one type of electric field meter dispenses with the probe and the usual sensor connecting leads. It uses small rod antennas which are plugged in as required. This instrument is in the form of a cube and the three antennas in a set are plugged into three adjacent faces of this cube.

Wide-band probes are increasingly used, the frequency coverage having increased with the use of modern manufacturing techniques. Two early papers (1972) on the use of thin film techniques in wideband probes were produced by Aslan[33] and Hopfer[34], respectively. A third paper (1980) on a wideband probe with a wider frequency band was produced by Hopfer and Adler[35]. These provide interesting information on the approach to probe design.

Frequency coverage

Individual commercial instruments are available covering power density, electric field and magnetic field measurement over portions of the spectrum up to 40 GHz. The basis is usually that of a meter unit capable of operating with two or more plug-in probes covering different frequency ranges, different full scale deflections and different measured quantities. The degree of versatility and the available probe combinations varies across different suppliers.

The utilisation of such instruments depends on the measurement commitments of the user. For example, an organisation using only HF transmitters will not need microwave measuring equipment. A microwave oven manufacturer may only be concerned with the microwave equipment needed to cover the frequency used in the ovens (usually 2450 MHz or 915 MHz).

Firms involved in the supply of transmitters for the whole of the communication and radar fields may require equipment covering all the spectrum from LF to the current limit of such instruments, 40 GHz. Indeed, they may be well ahead of this and using frequencies of 100 GHz and more, thus having to resort to more basic methods of measurement.

For those with a wide range of requirements the wideband power density measuring equipments are attractive, typically covering 0.3 GHz to 40 GHz or, in two cases, covering 300 kHz to 40 GHz. The prime attraction is, of course, not having to change instruments over this wide range which covers amongst other things, most radar equipments.

For measurement at the lower end of the frequency spectrum, various guides and standards (see Chapter 4) require the separate measurement of the electric and magnetic field components, power density not being considered to be a relevant parameter.

In the UK, the NRPB requirement applies below 30 MHz. In the USA, the IEEE C95.1–1991 provisions[4] require separate measurement of the electric and magnetic field components up to 300 MHz. This frequency is the current limit of availability of magnetic field measuring equipment.

Because separate probes are normally needed for electric and magnetic field measurement and these are designed for limited band coverage, the test equipment investment for the number of discrete probes which need to be purchased can be quite high.

A simplified diagram of the availability of probes against frequency for three manufacturers, limited to electric field probes for clarity, is given in Figure 6.6.

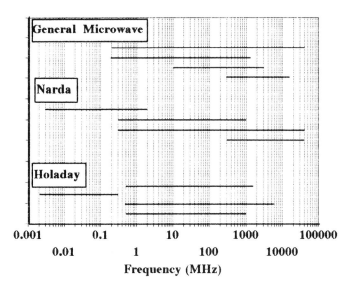

Figure 6.6 *Electric field probe availability from three manufacturers taken from sales literature*

It shows the typical coverage of the frequency spectrum with electric field probes and, in particular, the scope of the wideband probes. It should be noted that although single lines are shown there may be a number of probes represented by each line in the diagram, covering different probe sensitivities and maximum power ratings.

Specialised probes, including those specifically for microwave ovens, have been excluded from this diagram for simplicity. In the case of the Narda company which produces a very large number of different probes, the 87 – series only is shown as giving a reasonable picture of availability. There are nine electric field probes in this one instrument range.

Magnetic field probes are also available from two of the suppliers and can be used with the same basic instruments as the electric field probes.

At least one manufacturer now produces an instrument in the HF band and another in the VHF and UHF bands in which a single probe incorporates both electric and magnetic field sensors and the choice of measurement of these two is made at the throw of a switch. The dual arrangement of the electric and magnetic antennas in one probe is interesting and is illustrated in Figure 6.7.

Such instruments have the potential to reduce the burden involved in HF transmitter surveys where the risk of a mistake in remembering which of the separate probes need changing at which frequency is appreciable. One penalty is that the frequency range covered by each probe is not very large.

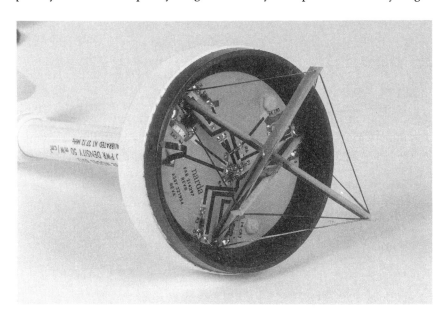

Figure 6.7   *Narda illustration showing a combined E and H field sensor in a single probe*
   (Courtesy Loral Microwave-Narda Company)

Measurement capabilities

Power density measurement

Power density is perhaps the most commonly measured quantity in RF safety management surveys. It is a common fact of life in the radio engineering field that we often require to measure a parameter which cannot conveniently be directly measured.

Although instruments are available which have scales calibrated in power density, such instruments do not actually measure power density, instead measuring either the electric or the magnetic field component. The electric field is used for the high frequency wideband probes for the reason given earlier.

It should be noted that technically the power density reading is not valid where the measured signal is not a plane wave, for example, in the near field. Instruments are calibrated on the assumption of a free space impedance of $377\,\Omega$, i.e. plane wave conditions.

Research to date into the possibility of designing a practical true power density measuring instrument for near field work has, so far not been successful. However, when existing types of instrument are used in the near field, then the electric field, for an instrument with a probe which measures that quantity, will be valid and can be retrieved by using the calibration equation in reverse. Similarly, where a magnetic field probe is used, the magnetic field value can be retrieved.

The scaling of RF radiation meters is a subject of some interest as it affects the amount of calculation to be done when using them. Power density instruments are mostly scaled in mW^{-2} although many people prefer Wm^{-2} in line with the majority of safety standards. It will be recalled that $1\,mWcm^{-2} = 10\,Wm^{-2}$, so that the conversion is easy to remember.

Hybrid units such as mWm^{-2} or Wcm^{-2} are a nuisance and can lead to safety problems due to erroneous interpretation. Hence $0.01\,Wm^{-2}$ can be safer than $10\,mWm^{-2}$!

Scales for power density are calibrated by the usual plane wave relationship:

Where an electric field probe is used, the calibration equation is:

power density S $(Wm^{-2}) = E^2/377\,\Omega$, where E is in Vm^{-1}

Where a magnetic field probe is used, the calibration equation is:

power density S $(Wm^{-2}) = H^2 \times 377\,\Omega$, where H is in Am^{-1}

Electric and magnetic field measurements

When using electric and magnetic field probes to measure the electric and magnetic fields using one meter unit and two probes, the scale is usually in power density, leaving the user to convert according to the probe used on the basis of the examples given above. In some cases, electric field scaling is in $(Vm^{-1})^2$ as given in the ANSI C95.1–1982 standard (but, happily, not in the IEEE C95.1–1991 standard).

In this case it is necessary to take the roots of the indicated values to obtain Vm^{-1} or to divide $(Vm^{-1})^2$ by $377\,\Omega$ to obtain plain wave equivalent power density. Working in the large numbers implicit in the $(Vm^{-1})^2$ scaling is another potential source of human error.

Where the meter is scaled in $mWcm^{-2}$ or Wm^{-2}, and an electric field probe is used, conversion to the electric field value uses the expression:

$$E\ (Vm^{-1}) = \sqrt{Wm^{-2} \times 377}$$

Where the meter is scaled in $mWcm^{-2}$ or Wm^{-2}, and a magnetic field probe is used, conversion to the electric field value uses the expression:

$$H\ (Am^{-1}) = \sqrt{\frac{Wm^{-2}}{377}}$$

Examples:

1 Electric field instrument

Instrument reading $2\,mWcm^{-2} = 20\,Wm^{-2}$

$$E\ (Vm^{-1}) = \sqrt{20 \times 377} = 86.83\,Vm^{-1}$$

2 Magnetic field instrument

Instrument reading $3\,mWcm^{-2} = 30\,Wm^{-2}$

$$H\ (Am^{-1}) = \sqrt{\frac{30}{377}} = 0.28\,Am^{-1}$$

Meter scale coding

One other aspect of scaling is of interest. Many instruments have a colour coding system for probes with the same colour used for the appropriate meter scale, for example, 'yellow probe: read yellow scale'. This takes into account instruments where a variety of probes can be used with one meter unit and it can be very useful in helping to reduce errors in reading instruments. Several years of training engineers and technicians in RF radiation safety has shown that even with such coding, misreading of scales is extremely common when people are first introduced to RF radiation measuring instruments.

Special instruments

One instrument, possibly unique, offers scaling both in conventional power density and also in terms of the percentage compliance with the ANSI C95.1–1982 limit values, such that exceeding a limit results in an indication of greater than 100%. This instrument is illustrated in Figure 6.8 later in this chapter. Although at first glance this does not seem to be a very remarkable thing to do, it is not just a matter of scaling. It requires that the ANSI limit values, which vary with frequency, be provided for by tailoring the frequency-sensitivity response appropriately, no easy task.

It can also be used when more than one signal is present since, providing that the frequency is within the instrument bandwidth, the signals will effectively be scaled against the relevant permitted limits and the addition of the signals will give a valid percentage of the RF protection guide as if this had been computed manually. Note that ANSI C95.1-1982 has now been superseded by the IEEE C95.1-1991 standard mentioned above and a probe has been developed for this standard.

Instrument features and limitations

Features

All electronic instrument design involves compromises, both technical and financial, and this information does not usually feature overtly in sales literature. However, some companies do, from time to time, issue technical papers which provide a little insight into the difficulties and achievements.

Also the USA IEEE produce a very useful document giving a recommended practice for RF field measurements, which includes discussion of instrument characteristics. This is IEEE C95.3-1991 [36]. This now combines and extends the coverage of the former ANSI documents ANSI C95.3-1973 and ANSI C95.5-1982.

There are a number of general features of field and other RF test instruments which have not been mentioned in detail in the earlier discussion of the basic operation of RF radiation measurement instruments:

1 Instrument zero

Some instruments have a single manual zero control or a coarse and fine ganged-potentiometer arrangement. Ganged controls can be a problem with wear so that the adjustment of one begins to disturb the other. Potentiometer knobs of any kind are easily disturbed accidentally with the hand unless they are guarded in some way.

Other instruments have a 'push to zero' button together with a screwdriver-adjustable coarse potentiometer. Some instruments claim to have fully automatic zeroing. Generally those instruments which are set to zero by a potentiometer or push button need to be taken out of the field to set zero but a few claim that zero can be set in the presence of a field.

Some caution should be exercised in selecting instruments with fully automatic zero, unless it can be established that the design is such that errors do not arise with amplitude modulated fields.

2 Alarm circuit

This is a valuable facility whereby an alarm can be set to operate at a preset fraction of full scale deflection. This is not only useful in making personnel aware of their situation but can also be helpful, if set appropriately, in avoiding damage to instruments due to overload by reminding the user to be ready to change ranges. It is often arranged that the alarm will sound continuously if the probe is burned out or otherwise becomes open-circuit with a cable break or connector disconnection.

3 Maximum hold facility

This is the facility to set the instrument to a memory mode where it stores the highest mean value of the field experienced during a survey. For example, in walking slowly through a microwave beam holding the instrument, the meter will indicate the highest value encountered and retain the reading when the user leaves the beam. Some manufacturers use the term 'peak hold' but this should be deprecated both because of the possible confusion with peak power in pulsed systems and because the instrument is not, in any event, a peak measuring device.

4 Space and time averaging

Some instruments provide facilities, either directly or by the use of an add-on unit, whereby readings can be averaged over a predetermined time limit for time averages, or averaged spatially, that is to say, averaged over a specific volume of space such as a room. For spatial averaging, the instrument is started and the surveyor then walks round in a pre-planned way at a speed slow enough to allow the instrument to register the field. At the end, the instrument is stopped and it calculates and displays the spatial average of the field measurements.

In the case of time averaging, there are various applications of which the simplest is to average a time-varying field over a defined period of time at a specific place. The instrument is set up to measure the field in question and put in a fixed location. Time averaging is initiated for six minutes or any other period available on the instrument. At the end of the time period, the instrument stops measuring and displays the time-averaged value.

5 Battery charging facility

Instruments use either dry batteries or rechargeable batteries. Those able to use the latter usually have a charger inbuilt. Rechargeable batteries do eventually need replacing, a fact that is often overlooked.

A battery state indicator is essential. One advantage of a rechargeable battery is that recharging can be done at the place where a survey is being undertaken, during breaks in surveying.

6 Choice of probes

Some instruments provide a wider range of choice than others. Not everyone will need a wide range and probes are expensive so the choices need to be thought out. Careful thought should be given to the CW and peak overload ratings when purchasing instruments, especially those to work with pulse transmission, in order to reduce the possibility of probe burn-out.

Some makers offer a choice of sensitivities and maximum ratings. This topic is normally addressed in equipment handbooks so that seeing the handbook before purchasing an instrument should be the rule! This will also identify probes which have not been designed for pulse work.

Where radiation surveys are a regular task, extra equipment may be needed to cover the loss of instruments during repairs and regular calibration. For large surveys, at least a two-man team will be necessary and some degree of duplication of instruments is therefore required.

When choosing instruments for a specific activity, a check list is often helpful and the following listing may provide a starting point. Trying out instruments before purchase can be essential unless the purchase is a repeat purchase. Similarly when new instruments are put on the market, evaluation is to be recommended.

All reputable suppliers recognise these needs and make suitable arrangements. The ergonomic aspects of instruments can be very important to some people whereas those who only use them for a few minutes at a time may not be much concerned. Try carrying the equipment which you propose to purchase for 2 or 3 hours if you need an instrument to conduct extensive surveys!

As with other electronic instruments, the advent of digital semiconductor circuits has resulted in the provision of many extra facilities in RF radiation instruments, some of which have been described above. The important thing, however, is what facilities are required rather than the number of facilities that are provided. Occasionally an instrument appears which lacks some essential facility. One from the past which comes to the author's mind is the lack of a battery voltage check on a particular instrument which used rechargeable batteries!

Limitations

The development of isotropic sensors has considerably improved measurement but since anything in production has imperfections or variations between items, any departures from true isotropicity give rise to a potential measurement uncertainty. Typical isotropicity uncertainties range from ±0.5 to ±1 dB. Variation of instrument response across the frequency band covered gives rise to a further uncertainty of measurement. This is given in specifications and obviously varies across the market but ±2 dB is probably typical for wideband probes.

Zero stability is also a source of possible measurement errors since if the zero setting is not stable it will be difficult to keep it correctly set. In consequence, the errors will be hidden in the measurements made. Also, if zero setting is inadvertently done with a field present, sooner or later a place will be found where that field disappears, giving rise to a mysterious negative reading. This applies to most instruments but there are some types where it is claimed that zero setting can be done in a field without giving rise to this problem. There are also one or two instruments which have automatic zero setting and this was subject to a note about possible errors earlier in this chapter.

Again, it is important that an electric field sensor should only respond to the electric field and not the magnetic field component and vice versa. However, in circumstances where the field being measured, say the magnetic field, is relatively small and there is a high level electric field present, then it is likely that the latter will affect the meter reading. This is largely a matter of correct instrument usage – avoid trying to do the impossible!

One trade-off involved in design is that of the length of the leads from the sensor to the amplifier or instrument unit. At very low frequencies, (<1 MHz) the leads may act as an extension of the 'antenna', picking up RF and giving erroneous readings. In some cases, low frequency probes may incorporate active sensors with an amplifier preceding the detector, thus reducing the significance of the pick-up from these leads.

Calibration also poses a basic limitation since the uncertainty of measurement obtainable at a calibration laboratory is a determinant in the fixing of the instrument uncertainties. The availability of suitable calibration laboratories across the world is both variable in distribution and limited in number. The major instrument manufacturers have good calibration facilities and supply calibration correction factors with new instruments.

A different type of problem is that mentioned earlier of the erroneous readings which may be obtained if there is a source of strong signals outside the bandwidth of the instrument. If these produce a reading, it may be taken as being a valid reading for the field currently being measured.

Realistically, it has to be recognised that as with any other type of frequency sensitive filter, out-of-band frequencies are just attenuated and if the amplitude of any such signal is sufficiently large, it will make its presence felt as a false reading. If there is any suspicion of this, switching off the wanted source may confirm that an out-of-band signal is present. It is necessary to be alert to the possibility that the source switched off was also the source of the out-of-band signal!

The presence of such signals may be predictable from a knowledge of the area but the amplitude may not be predictable. The onus is therefore on the user to exercise care. The minimisation of such problems requires judgement on the part of the user and might involve the use of other instruments such as a spectrum analyser to identify large out-of-band signals.

Knowledge of the performance of instruments with respect to out-of-band responses depends largely on the information supplied with the instrument.

This is not always as much as might be desired! Where out-of-band responses are obtained with relatively low amplitude signals, this is a cause for concern and the instrument needs further investigation. It is not unknown for a particular model of instrument to have some out-of-band weaknesses over particular frequency ranges which can only be found by systematic evaluation.

At lower frequencies, below about 10 MHz, investigation has shown that differences in electric potential between the probe and the meter unit, for instruments having high resistance sensor output connections, results in erroneous readings due to the instrument responding to the potential difference. The errors in readings are said to be greatest at the low frequencies used for the AM broadcast band[37]. It may not apply to those where the antenna is integral to the meter unit and does not need to have high resistance connecting leads.

It can be seen from these examples that specific measurement results will not only be affected by instrument calibration and the stability of instrument characteristics between calibrations, but also by factors in practical use which result from the limitations of instruments and the judgements of the user.

As noted earlier, RF radiation meter sensors, diode or thermocouple are very vulnerable to overload and burn-out of the elements. Users often forget that the sensor system is there to pick up energy and is not made safe by merely switching off the instrument. It is easy to forget this when walking about with an instrument. Hence probes should be kept in the carrying case or covered with a shield such as aluminium cooking foil.

When working with pulsed transmitters or RF machines, it is important to check the pulse duty factor against the maker's operating instruction manual, since some limitation of reading may be needed to avoid damage at low pulse duty factors (see Chapter 8). It is still common to find that when no reading is obtained near a waveguide aperture, engineers will, if it is at all possible, push the probe end into the aperture and destroy the sensor before they realise what is happening.

The replacement of probes is a very expensive matter. Whilst some manufacturers are looking at methods of reducing this cost, there is also a loss of utilisation whilst repairs are done, and this can sometimes be considerable. In the author's experience, most such instruments are very reliable but the number of times the sensor is burned out seems to be proportional to the number of people allowed to use the equipment and may perhaps owe something to a lack of personnel training.

Technical checklist for instruments

The list below is not arranged in any particular order. The order of priority for the various choices will depend on the needs of the user.

1 Frequencies to be covered and the frequency coverage of the instruments being considered.
2 Choice of probes: diode or thermocouple sensor detector, where there is a choice. Suitability for pulse and amplitude modulated signals (if required). For diode types, check the maker's provision to reduce the measurement errors referred to earlier.
3 Dynamic range.
4 Overload safety margin, especially for pulsed modulation.
5 Quantities to be measured: Wm^{-2}; Vm^{-1}; Am^{-1}.
6 Meter scaling.
7 Required operating temperature and humidity range.
8 Investigate protection against out-of-band frequencies carefully!
9 General facilities required ('needed' rather than 'nice'): maximum hold; alarm; space/time averaging, etc.
10 Mass and ergonomics.
11 Knowledge of reliability and durability.
12 Options e.g. optical fibre connection, etc.
13 Battery options.

Practical examples of RF radiation instruments

Field survey instruments

It is only possible to use a limited number of examples in this section as the main instrument producers between them have a very large number of instruments and probes in their catalogues. An attempt has been made to use specimens which illustrate some specific aspects of modern RF radiation instruments. Some common features are discussed towards the end of this section.

The instruments illustrated here are typical ones and many, but not all of them, have been used by the author. However any reference here to an instrument is solely intended to highlight features and does not imply any recommendation. Equally, there is no adverse implication in the absence of any instrument manufacturer or instrument.

The actual choice of instruments should be made against the specific requirements of the work and the fields to be measured. The instrument specification data should be checked from the supplier's data sheets since these change from time to time.

Figure 6.8 shows the Narda type 8716 meter unit. The probe also illustrated (type 8721) is an electric field isotropic probe with a flat frequency response covering 300 MHz to 40 GHz. Alternatively, the type

Figure 6.8   *Narda ANSI conformal probe type 8716 (Courtesy Loral Microwave-Narda Company)*

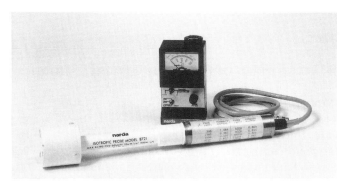

Figure 6.9   *Narda type 8711 meter unit and probe (Courtesy Loral Microwave-Narda Company)*

8722 electric field isotropic probe which has an ANSI standard C95.1-1982 shaped response, may be used. This ANSI conformal probe was mentioned earlier in this chapter and covers the frequency range 300 kHz to 40 GHz. It utilises a meter unit scaling which indicates in terms of the percentage conformance to that standard (100% = ANSI limit).

The nomenclature used for the ANSI response probe is '% RFPG' where this is an abbreviation for percentage of the 'RF Protection Guide' ANSI C95.1–1982. The meter scale for this is the top scale and on the range switch the segment below the switch gives switched scales of 3, 30, and 300% of the standard. It also has the conventional scaling to operate with normal flat response probes and is compatible with a number of 8700 series probes, both magnetic and electric.

An interesting feature of the meter unit is that it contains built-in test sources and probes have contacts on the side which can be held in electrical contact with studs on the side of the meter unit case which connect to the low frequency source output. There is also a microwave slot radiating test source (12 GHz) against which the probe can be checked. With the aid of these, probe functionality can be tested without any other equipment being needed.

The other meter scales are conventional $mWcm^{-2}$ ranges for flat response probes from 0.2 to 2 000 $mWcm^{-2}$ (2 to 20 000 Wm^{-2}). Above the range switch, coloured segments relate different probes to their related scales. The meter unit has the usual facilities, an adjustable alarm, maximum hold switch and automatic zeroing. The mass of the meter unit is 2.1 kg (4.5 lb.). An alternative meter unit (model 8711) can be used with the 8700 series probes and which has a mass of only 0.6 kg (1.3 lb.) but, of course, offers less facilities. It could be described as pocket-sized and lends itself to slipping in a coverall garment top pocket when climbing towers and structures. The instrument is shown in Figure 6.9. The meter unit size is 118 × 67 × 45 mm (4.63 × 2.63 × 1.75 in.) against 228 × 135 × 105 mm (9.6 × 5.6 × 4.4 in.) for the one in Figure 6.8.

An optional unit which can be used with this instrument is the 8696 radiation averaging module which provides time and spatial averaging and reads in percentage of full scale for the range selected on the main instrument. It is shown in Figure 6.8, fitted on the top of the instrument. It is referred to again in Chapter 9.

Figure 6.10 shows the General Microwave Raham model 30 RF radiation meter with an isotropic probe covering 300 GHz to 18 GHz. The power density range covers 0.2 to 200 mW^{-2} (2 to 2000 Wm^{-2}). Facilities include time averaging over 6 minutes, maximum hold and an alarm. The meter scaling is in $mWcm^{-2}$.

The instrument is held by means of a rear handle which can be seen acting as a rear stand in the picture. The mass of the meter and probe is 1 kg (2.2 lb) and is thus quite light to handle. The probes may be plugged rigidly into the

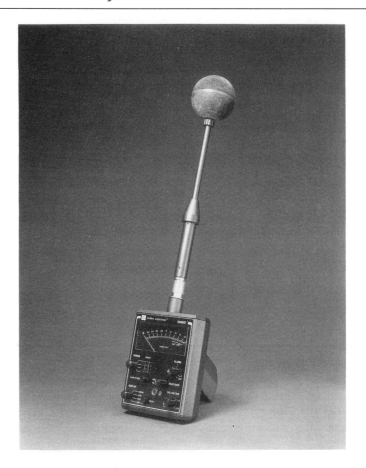

Figure 6.10   *General Microwave Raham model 30 radiation meter and probe type*
   (Courtesy General Microwave Corporation)

socket, or an extension cable can be used to improve manipulation in awkward situations. Model 40 has a similar appearance but covers 200 kHz to 40 GHz. The measurement range is 0.002 to 20 mWcm^{-2} (0.02 to 200 Wm^{-2}). A paper provided by General Microwave Inc. [38] gives details of the construction of their instruments. It includes a reference to the issue of diode linearity in those probes which use them and to those situations where errors can occur. Similarly, details of the use of thermocouples in some of their probes are given.

Figure 6.11 shows the Holaday model HI-3012 instrument with the electric field probe fitted. The instrument is an isotropic broadband field

Figure 6.11 *Holaday model H-3012 radiation meter and probes (Courtesy Holaday Company)*

strength meter. An electric field probe and a magnetic field probe are supplied as standard. The magnetic field covers 5 MHz to 300 MHz and the electric field probe 0.5 MHz to 5 GHz. The meter has an automatic zeroing facility which operates even when carrying out measurements. It also has the usual maximum hold facility. The meter is scaled in 'field strength units squared' marked as $(FSU)^2$, i.e., E^2 or H^2. These are converted as required by the relevant formula.

The instrument is sensitive enough to measure the lower 'uncontrolled area' levels specified in the IEEE standard C95.1-1991. The effective ranges of measurement expressed in equivalent plane wave terms are: electric field 0.3 to 2650 Wm^{-2}; magnetic field 0.4 to 3770 Wm^{-2}.

A separate small unit, the HI-3320 data logger, is an add-on unit which provides for 0.1 hour (6 minutes) averaging of measurements, where required.

Another approach to electric field measurement is the Instruments for Industry Inc., (IFI) radiation hazard meter RHM1 shown in Figure 6.12 (the larger instrument). This instrument has a cube-shaped case with three

Figure 6.12   *RHM1 and EFS1 electric field meters (Courtesy Instruments for Industry (IFI) Company)*

antennas, one fitted on the top, one on the side and one on the back face of the case so that they are orthogonal. The meter is scaled in volts per metre and the range is changed by plugging in an alternative set of antennas. The range is determined by the length of the antennas in each set. It is possible to switch off any of the antenna 'channels' with front panel switches so that one, two or three antennas can be in operation.

The direct insertion of the antennas dispenses with the need for any significant length of connecting leads. However, the whole instrument with antennas fitted constitutes the equivalent of a probe and is substantially larger than the usual probes. An optional dielectric handle can be fitted to the unit to allow it to be used as a probing device. The frequency range is from 10 kHz to 220 MHz and the signal range is 1 to 300 Vm^{-1}.

The unit uses a rechargeable battery. The type RHM2 version includes a light modulator transmitter to drive a fibre optics link.

The smaller instrument shown alongside the RHM1 is the EFS1 which has a single antenna and is capable of measuring the peak value of the electric field for CW and pulsed transmissions, with some limitations on the latter.

The equipments mentioned above are general purpose. However, Narda and Holaday also make a number of instruments for specific applications involving the microwave oven frequencies of 2450 and 915 MHz.

With the odd exception, the conventional instruments described above do not go below 200 or 300 kHz. The exceptions are those electrical field meters made by IFI which go down to 10 kHz. There are broadcast transmitters down to something like 150 kHz and RF machines and some transmitters well below this figure.

Some instruments are becoming available in the lower frequency ranges. One such is the Holaday HI-3603 which covers 2 to 300 kHz electric field and 8 to 300 kHz for the magnetic field. The electric field measurement range is 1 to 2000 Vm^{-1} and the magnetic field range is 1 to 2000 mAm^{-1}. The equipment is primarily aimed at visual display unit (VDU) measurements but may have some application in the field mentioned above.

There is now an increasing amount of equipment being developed in these lower frequency ranges, predominantly for the general EMC measurement field. In the latter case the levels measured are those typical of EMC rather than the higher levels reflected in RF safety standards. Nevertheless there may be some application for such instruments in RF safety work. Consequently, most of the manufacturers mentioned in this chapter are also beginning to develop products aimed at the EMC market.

Diagnostics

In the context of X-ray measurement the concept of a 'sniffer' instrument is quite common. It applies to a simple and sensitive instrument used solely to locate leaks, leaving formal measurement to other instruments.

Various types of RF sniffer devices come on the market from time to time in the EMC field. These are generally used for checking for unwanted and usually low level RF radiation from products, i.e. leakage which could cause interference. The author has not until recently come across such devices which were intended for, and safe to use with, the sort of levels found in transmitter leakages.

There are perhaps two reasons why such a device could be useful in the latter category:

1 When undertaking leakage measurements with standard RF radiation instruments, even allowing for the variation between different brands of instrument, it can be difficult to get probes into small apertures and close

to waveguide flanges. Also, unless the meter unit can be propped up in some way, it is likely to be a two-handed activity.

2 When investigating the inside of high voltage transmitters, there may be potential dangers of electrocution.

The Narda Company have recently produced a slim insulated flexible probe, model 8781, to meet this need. The frequency range is 2 to 18 GHz and the probe is isotropic. However, it goes further than the simple requirement mentioned above since it offers two options. It can be used with a conventional meter unit such as the model 8716 to measure power density.

Alternatively, it can be used with the small audible frequency generator, model 8710 (50 to 5000 Hz) to give an indication where the frequency of the tone increases as the leak field increases when the probe gets nearer to the direction and source of the leak. Earphones are provided for noisy environments as transmitting stations are usually noisy due to air and coolant systems. The probe is shown in Figure 6.13, fitted with the audible frequency generator.

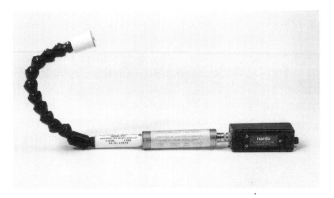

Figure 6.13   *Narda flexible probe type 8781 with audio tone generator 8710*
(Courtesy Loral Microwave-Narda Company)

Systems for laboratory and production use

Apart from portable instruments there are some instruments which perhaps best come under the heading of systems.

The Electro-Mechanics Company produce a system which comprises a data processing and interface unit which can be connected to up to eight electric field probes, each of which has a separate meter unit with an optical fibre output to a data processing/interface unit.

The probes are all electric field sensors and are mounted on stands. Probe coverage is 10 kHz to 1 GHz for two of the probes and 10 MHz to 18 GHz for the third type. The system can be set up in a specific configuration around a transmitter, RF machine or medical RF generator and the sensor readings logged.

Another type of facility which might be sought by those involved in developing automatic testing equipment (ATE) for electronic products, is a system which could lend itself to use with, or be incorporated in, an ATE system. A Narda system comprising a unit which interfaces with a computer can, when coupled with the selection of a suitable probe for the purpose in hand, be arranged to carry out measurements on the unit under test.

Similarly, Holaday manufacture, on a custom basis, automatic testing systems for microwave oven manufacture. Fixtures hold the oven under test and an array of probes makes the required leakage tests and prints out the results. The degree of automation used is determined by the purchaser's requirements. Figure 6.14 illustrates one such system.

Figure 6.14   *Holaday automatic testing installation for microwave ovens (Courtesy Holaday Company)*

Here there are two stations for the test sequence, the first performing test leakage scans from the oven back and bottom whilst the second station does the full scan test on the oven front, door and seals. A powered conveyor system is used for movement of the ovens under test. In a complete system, a quality assurance audit station and a repair station is included. The test data is printed out and also passed to disk storage.

With the increased emphasis being placed worldwide on the electromagnetic compatibility (EMC) testing of products and increase in safety testing of transmitters and RF machines there is likely to be a trend towards the use of automatic testing equipment for both these types of test, which have some basic common features in terms of the equipment used.

Personnel safety monitors

There has long been a need for some sort of small and low cost device capable of either being worn, or of being located near a potential source of RF radiation.

Comparison has tended to be made with the film or thermo-luminescent dosemeters used for ionising radiation. So far nothing with the simplicity of these badges has been devised for RF, though attempts have been made using liquid crystal materials and other techniques.

There have been a variety of small low cost diode-based RF radiation indicator devices around the world, many of which have been the subject of adverse evaluations by interested bodies. It does not appear possible to produce anything which has a reliable measurement capability at the low cost envisaged by many potential users.

Indeed the design of a reasonably wideband pocket-size personal monitor is an expert and expensive task. Consequently, very low-cost instruments claiming to offer such a performance should be treated with considerable caution and more than one sample evaluated in practical situations. The usual problem is the unpredictable variation of characteristics exhibited by the low cost devices so that unsafe leaks may be indicated as safe and small acceptable leaks indicated as hazardous.

The desire for a personal monitor stems from those who work in an environment where they may become unknowingly exposed to RF due to some unpredictable event in their work pattern. A good example of this is work on the servicing and fault-finding of equipment which produces appreciable RF such as military aircraft with radar and countermeasures equipment where the equipment may be in the aircraft or in a detachable pod. Civil aircraft and ship equipment maintenance may also have this sort of requirement.

A similar need may arise in the first testing of newly assembled microwave transmitting equipment where possible waveguide faults or flange couplings may throw up leakages.

The basic requirement, leaving aside the inevitable personal desire for embellishments, is to be alerted when a preset threshold power density (or other field quantity) is exceeded.

To cope with noisy environments a latched indicator (one which remains on until reset) is postulated so that if the audio alarm is not heard, the light can be seen and provides a record of what has happened. An alternative or additional facility is the provision of a simple earphone.

The basis of the use of such devices is that, should the monitor indicate an alarm condition, somebody has to be available with measuring equipment to investigate the problem.

The technology involved is basically the same as that described for other RF radiation instruments but the significant physical differences are:

1 A relatively simple range of facilities are needed compared with measuring instruments of the type illustrated earlier in this chapter.
2 Size is a major parameter as the device is attached to clothing.
3 The monitor is not usually expected to have a measurement read-out but only to provide an alarm.
4 The mass has to be small and there is a problem in relation to possible battery capacities and hence the utilisation time between battery changes.

There is still a tendency to ask for more and more facilities once a basic instrument becomes available. Whilst this is not unusual, it may not be desirable to stretch monitors into full measuring instruments!

Figure 6.15 shows a typical monitor of the personal type intended for wearing at work and manufactured by the Narda Company. Model 8841C covers the frequency range of 1 to 18 GHz and the model 8840C 2 to 18 GHz. Both use a thermocouple detector built into the sensor antenna structure.

The back of the instrument, that is to say the face which is in contact with the body, has a layer of RF absorber inside to reduce the effect of energy reflected from the wearer's body. The half power beamwidth is 120° in the vertical and horizontal planes in front of the wearer. The size of the device is roughly that of a pack of cigarettes and it has a clip to attach the instrument to the user's clothing.

Model 8842C is available with the frequency range 0.8 to 6 GHz, which covers, amongst other things, the mobile telephone field where service engineers have to work on masts or mounts which may have a proliferation of antennas.

Another model (model 8843C) is under development covering 30 to 600 MHz which covers the VHF/UHF bands. This will be relevant to the broadcasting field and particularly to the problems of antenna tower climbing. Model 8844C covers the HF band from 2 to 30 MHz.

These monitors have a lamp indicator as well as an audible alarm. The level at which the alarm operates is preset at either $10\,\text{Wm}^{-2}$ or $50\,\text{Wm}^{-2}$

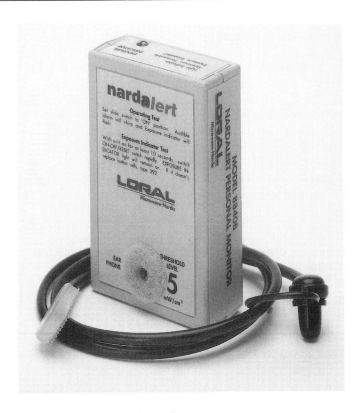

Figure 6.15   *Narda personal monitor
(Courtesy Loral Microwave-Narda Company)*

Figure 6.16   *GMC personal monitor
(Courtesy of General Microwave Corporation)*

(1 or 5 mWcm^{-2}). The audible alarm beeps at a rate which increases with the field level. The lamp latches on after the preset operate level has been reached and the light remains latched until manually reset. Functional self-checking takes place on start up and battery status is monitored. Alarm reset is performed by a single switch which is also the on-off switch.

Figure 6.16 shows another of these devices, this one being produced by General Microwave Corporation (GMC). It also uses a thermocouple system and covers the frequency range from 1 to 18 GHz. There are two versions of the type 65 (65-1 and 65-5) to cover the two preset alarm levels 10 Wm^{-2} or 50 Wm^{-2}.

When the preset latching alarm level is reached, the unit bleeps and flashes and the bleep and flash rate are proportional to the RF field strength. Self-checking is in the form of continuous monitoring of the detector and the battery. Instrument or battery failure is indicated by visual and audible signals.

The model 60 instrument allows the alarm level to be set by the user between 2 and 200 Wm^{-2} and also provides a measurement mode and a 6 minute averaging facility. In the measurement mode, the instantaneous exposure level is indicated on a liquid crystal display.

Both models are fitted with a clip for attachment to clothing and the size is similar to the Narda one. Both the Narda and GMC types offer compact monitors which are easy to operate and wear.

An alternative approach to the personal monitor is the general personnel monitor. With this arrangement, one or more sensors are installed in fixed locations within a room, cabin or other working space, thus protecting those people who may have access to it.

Philosophically, the argument in favour of such systems is that protection is automatically afforded to anyone, whereas personal monitors are individually allocated and therefore only protect those who possess them and remember to wear them. The main limitation of such devices is that they need to be located near any likely leakage.

It is sometimes thought that it ought to be possible to fit a single device somewhere in a room, rather as smoke detectors are fitted, and rely on this to provide an audible warning.

However, leaks from equipment which involve hazardous levels do not generally extend far in space, unless the leakage aperture is large. This means that personnel could work in a potentially hazardous field without it operating the sensor unless the sensor is located close to the source of the leakage. It may be that more than one sensor is needed in such cases.

Where fixed systems are appropriate, they can provide for a degree of sophistication in the safety provisions when associated with other sensors e.g. infrared sensors for the detection of the presence of people or by the use of the RF detector alarm signal to operate illuminated warning signs or prevent access by operating electric locks.

Narda produce a number of battery-operated sensors which can be used for this purpose. They go under the name 'Smarts'. One is illustrated in Figure 6.17. The Smarts devices cover a range of frequencies. Model 8810

Figure 6.17 Narda 'Smarts' radiation detector
(Courtesy Loral Microwave-Narda Company)

covers 2 to 30 MHz, model 8815 covers 10 to 500 MHz and model 8820 covers the frequency range 0.5 to 18 GHz. Each provides an aural and visual alarm plus a DC output signal which can be used locally to operate other warning devices. There is also a facility to use an optional infrared personnel detector so that the alarm can be inhibited when no personnel are detected in the area. The device is battery operated and includes the usual low battery indication. The factory set alarm level is $10 \, \text{Wm}^{-2}$. The supplier's instructions address comprehensively the problem of siting and installing the devices.

Holaday similarly produce monitoring equipment which can be located in a suitable place and which can provide a contact closure at a preset field level to operate warnings or to shut down a transmitter. The HI-3500 is a hazard monitor which can be mounted in workplaces and which can use any Holaday isotropic wideband probe. The frequency range is thus determined by the choice of probe. Apart from the contact closure output, there is an optional optical fibre link remote read-out.

Holaday also produce several similar devices which are for use at 2450 MHz, the usual microwave oven frequency.

Induced body and contact current measuring systems

Induced body current monitors

With the new emphasis on induced body currents and the specification of limits for limb currents in standards, there is a need for instrumentation capable of measuring these currents. The basic method for measuring induced body currents is illustrated in Figure 6.18. This shows a person standing on a metal plate which itself is mounted on an insulating plinth.

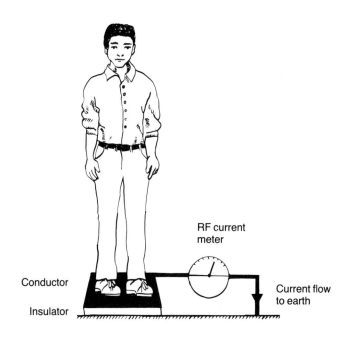

RF current
meter

Conductor

Current flow
to earth

Insulator

Figure 6.18 *Induced body current measurement concept*

The plate is shunted by a low resistance through which the current can flow to earth. The RF voltage across this known resistance is measured and the current calculated.

The practical realisation of this is clearly somewhat better than the basic illustration. However, because of the compactness of practical instruments, they do not really show much if used for illustration. The objective is to minimise the length of the conductors shown connecting the meter in Figure 6.18 since otherwise they may provide direct induction of current.

Another factor is the fact that a thick plinth effectively increases the height of a subject. Also the plate to ground capacitance is important, as is the change of reactance of that capacitance with frequency. Most research work on this subject is done with the subject bare-footed since this is the worst case. The effect of the use of shoes has also been investigated as noted earlier.

From the early experimental work in this field, notably by Gandhi, at least two manufacturers – Narda and Holaday – have developed equipment and are developing further devices. Since this is a new field, the instruments available are described in some detail. Where instruments are still in development, there is the possibility of changes in the specifications but the information given here will provide an impression of the trends in this expanding field. Work has also been done by the UK NRPB on this subject.

The two firms have so far publicised involvement in the development of the following equipment for this purpose, some of which is now available.

The Narda Company

The Narda model 8854 induced current meter 'I MAT' is a thin mat intended to be placed under the feet of the operator of an RF process machine or a technician working on transmitter equipment in the frequency range of the equipment (50 kHz to 45 MHz).

When used in a working situation, the operator or technician has only to stand on the mat and is otherwise not disturbed as far as the work in hand is concerned. The instrument has two current ranges, 23 to 63 mA and 60 to 200 mA. The audible alarm level is preset and will sound when the induced current through the feet to earth reaches that setting. A recorder output is available for use in longer term observations. The 'I MAT' is powered by AC mains.

The model 8850 induced body current meter is still in development. The preliminary specification covers 3 kHz to 100 MHz and the current ranges are 10, 100 and 1000 mA full scale together with a range full scale calibrated as 250% of the IEEE C95.1-1991 standard for limb current. The instrument is powered with rechargeable batteries. It consists of a small plinth roughly 300 × 300 mm (12 × 12 in) on which a person stands. It will provide a more accurate measurement than the 'I MAT'.

An interesting feature is that a 'standard man equivalent' antenna is also being developed so that measurements can be made with a passive standard man instead of a human being. The plinth has a calibration factor control to allow for the fact that the 'standard man' is difficult to make and it is thought likely that it will be about 80% of a standard man in terms of body current measurements. The instrument is intended for use by the technically competent whereas the 'I MAT' is intended for use by anyone.

The Holaday Company

Holaday produce the model HI-3701 induced body current meter on a similar principle. From the sales literature, the appearance is almost identical to a slim, neat, bathroom scale. The dimensions are 360 × 485 × 50 mm (14 × 19 × 2 in) It has a current meter where the weight indicator would normally appear and a range switch alongside. It is powered with rechargeable batteries. As with the other types, it is used by standing on it.

The frequency range is 3 kHz to 100 MHz. The current ranges are 3, 10, 30, 100, 300 mA full scale. A recorder output is available and an optical fibre read-out is an option. The instrument does require the user to read the currents from a meter, which makes it a little more obtrusive as far as users engaged in practical work are concerned. On the other hand direct reading is ideal for some types of research.

The National Radiation Protection Board (UK)

The NRPB have been carrying out research using a different approach to the measurement of limb currents[39]. The argument for looking for an alternative method involved consideration of the possible shortcomings of the plinth approach discussed above. The points advanced were that the metal plate on which the subject stands can affect the fields, the plinth will raise the height of the subject and the RF impedance from the plate to an effective ground connection is put in series with the body of the subject.

Whether, in practice, these problems are significant is hard to say, especially as the only present method of cross checking is against results obtained from use of the plinth method! However another point made by the NRPB which does have a clear significance is the recognition that standing on a plate is somewhat limiting for a person carrying out work which requires mobility.

Consequently the NRPB approach was to produce a device which could be clamped on to the ankle. This used a current transformer and electronic circuitry with a liquid crystal display all fitted in situ so that the wearer can walk about. It is referred to as a personal current monitor (PCM). The report claims that an experimental model performed well over 100 kHz to 80 MHz and 8 to 1000 mA.

Comparisons up to 30 MHz with the plinth method but using an RF thermo-ammeter and the PCM device were shown as giving good agreement on plinth to ground currents measured with the thermo-ammeter and the PCM device used round the grounding braid.

Above 30 MHz there was not good agreement between PCM ankle current measurements and plinth current measurements. Also, in high field strengths it was found that the early experimental model was liable to give spurious readings due to interference with the PCM electronics.

Contact current measuring meters

There is also a requirement to provide for a method of measurement of contact currents such as those which may flow when contact is made with metal objects located in an RF field. It is often the case that metal masses are found in places which are irradiated by RF fields and relatively close to the radiating antennas.

Sometimes this may be in the form of spare antenna structures, towers, tower parts, vehicles and similar items. The measuring requirement is to ascertain the current which flows when a person touches such a conductive object with a view to avoiding any possibility of burns resulting.

The most basic method of measuring these currents is shown in Figure 6.19. The method involves the use of an RF thermo-ammeter. The subject

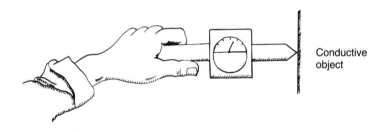

Conductive object

Figure 6.19 *Contact current measurement concept*

holds one terminal and touches the other on the metal object. In the illustration the person actually holds a substantial conductor to reduce the possibility of RF burns. The other end of the meter is shown as a pointed conductor with which to touch the metal object.

Apart from the technical limitations of this type of meter, it does, in principle, work for hand contact currents. Modern instruments can be made with an ordinary meter and suitable electronic circuitry to give a direct reading of RF current.

Narda model 8870 contact current meter is intended to simulate the contact current which would be experienced by a person with bare feet which are grounded, making a point contact with metal objects in the RF field. The instrument has a current range of 0 to 2000 mA with a facility to switch to a full scale range of 1000% of the IEEE C95.1-1991 standard. The frequency range is 3 kHz to 30 MHz.

Note that commercial instruments being developed use a man-equivalent circuit rather than allowing current to flow through the user!

Ancillary equipment

Occasionally, equipment such as spectrum analysers and measuring receivers may be necessary, particularly when tracing unexplained readings or when EMC investigations are also included in surveys.

These may also be needed when radiation levels are too low to measure with normal radiation equipment but quantitative information is needed for legal or planning purposes or just to convince those who insist on figures!

There is a vast amount of such equipment available worldwide if only because of the increase in interest in ensuring that new products do not cause interference and are not unduly susceptible to interference. This interest applies to military products where it is not new and now, increasingly, to industrial and domestic products as exemplified by the current EEC Directive on EMC[68].

In some leakage testing on high power broadcasting transmitters, it is sometimes possible to pick up unwanted signals such as very low frequency control signals from the transmitter which provide out-of-band readings on RF radiation instruments which are indistinguishable from the transmitter RF output. In these situations some sort of selective measuring instrument or a spectrum analyser which is suitable for the frequency concerned, may be needed to identify the signal.

With the advent of standards such as the INIRC 1988 and IEEE 1991 standards which introduce lower field levels for public exposures, 'uncontrolled environments' in the terminology of the latter standard, measuring equipment which is usually associated with EMC measurement may sometimes have some uses when measuring very low levels.

7
X-rays and X-ray measuring instruments

The nature of X-ray radiation

Most people are aware of the use of X-rays in medical diagnostic work and may, indeed, have had practical experience of it. Whilst it provides benefits of a unique kind in this application, it has a nuisance value when produced in situations where it is not wanted. Since X-ray production is associated with electrons being accelerated in a vacuum, the various types of high voltage electronic tubes used in radio equipment are potential sources. The applications include transmitters, some RF process machines, and cathode ray tube-based equipment such as television sets and visual display units.

Some knowledge of the nature and hazards of X-rays is necessary for those working with RF radiation since the two forms of radiation may co-exist. X-rays and gamma radiation involve electromagnetic waves similar in nature to RF radiation, but having wavelengths much shorter than those at the highest end of the radio frequency spectrum and are classified as ionising radiations.

X-rays and gamma rays are emitted as quanta or 'packets' of electromagnetic radiation, generally referred to as photons. These two forms are identical in nature but have different origins. Gamma radiation is emitted as a result of changes in the nucleus of substances known as radioactive substances and has no relevance here. X-rays result from the stopping of electrons, for example, by the anode of an electronic tube, after being accelerated by high voltages. As with radio frequency radiation, X-rays can be reflected and this is often important when considering the design of equipment cabinets.

It follows that 'solid-state' transmitters (transmitters using semi-conductor devices instead of electronic tubes) using the normal low supply voltages do not generate X-rays unless some high voltage is developed by conversion of the low voltage supply to operate electronic tubes such as cathode ray tubes and similar devices.

The wavelength of X-rays is several orders smaller ('shorter') than the wavelengths used for radio and radar purposes, as was shown earlier in Chapter 1, Figure 1.8. The energy of the radiation is inversely proportional to the wavelength, so that short wavelengths have higher energies than the longer wave radiations.

The term energy has been used above and this needs some explanation. The SI unit of energy is the joule (J). A more convenient unit, the electron-volt (eV), is used in ionising radiation work. It is defined as the energy acquired by an electron when accelerated through a potential difference of one volt. Hence, to take a common example, a cathode-ray tube with an anode potential of 12 kV will impart an energy of 12 keV to an electron traversing this potential difference. 1 eV is a small quantity, being equal to 1.6×10^{-19} J.

An idea of the wavelength for different energy levels is given in Figure 7.1. Note that in books wavelengths for X-rays are often found expressed in

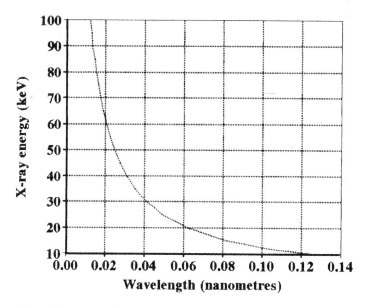

Figure 7.1 *X-ray wavelength versus tube voltage*

angstroms (1 Å = 10^{-10} metre). Frequency is rarely used in connection with X-rays but can, of course, be calculated as for radio frequency wavelengths.

The X-ray wavelength (λ) in nanometres can be calculated for a given photon energy as follows:

λ(nm) = 1241/photon energy (eV)

Example: 10keV gives a wavelength of 0.124 nm

Note that in the expression above, the 'e' is often omitted, particularly in data from the medical X-ray field, and tube voltage is used instead, i.e., V instead of eV, so that a 10 keV and a 10 kV accelerator are the same thing. These alternatives do not make any difference to the calculations.

The most ready source of information on X-ray production in electronic tubes comes, not surprisingly, from the manufacturers of X-ray tubes for use in medical diagnostic equipment. Here the aim is to produce X-rays, whereas in radio applications the X- rays are a nuisance. Figure 7.2 gives a

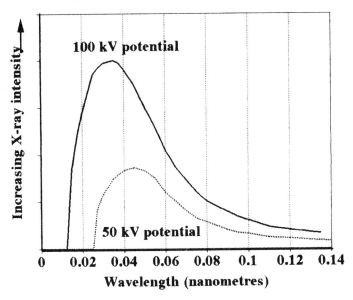

Figure 7.2 *X-ray energy distribution versus tube voltage*

general idea of the X-ray energy distribution and wavelength for potentials of 250 kV and 50 kV.

It can be seen that each spectrum is continuous to a specific short wavelength limit which is determined by the peak voltage across the tube. The actual shape of the energy distribution will depend on the nature of the applied tube voltage, for example, constant or pulsed. The long wave end of the spectrum is not sharply defined, some of the long wave energy being stopped by the glass of the tube, electrode structure and any filtration included.

The peak energy in the distribution occurs at about two or three times the shortest wavelength. Within the spectrum there will also be a narrow band or bands of energy specific to the tube anode material. This is not identified in Figure 7.2.

The effect of increasing the applied voltage is twofold. The short wave limit of the spectrum is decreased in wavelength and the intensities of all the wavelengths present are increased.

If the tube current (the beam current) is increased, the intensities of all the wavelengths present increase proportionately. Note that X-radiation doserate (which is discussed later) is proportional to the X-ray intensity.

In radio transmitters and similar equipment the peak accelerating voltage experienced by electrons in the final power amplifier tube will not correspond to the transmitter supply voltage but may be two or three times that voltage and this must be borne in mind when designing equipment.

Low energy X-rays are generally referred to as 'soft' X-rays and higher energy X-rays as 'hard' X-rays. In turn these are inclined to be shortened in conversation to 'soft' and 'hard' radiation. When referring to wavelengths, high energy X-rays are also referred to as 'short wave' radiation and the low energy end of the spectrum as 'long wave' radiation. These should not in any way be confused with the radio broadcasting usage of the terms.

From the point of view of the transmitter engineer, the energy of the X-rays is an indicator of the penetrating power in materials, higher energy X-rays requiring thicker shielding for a given material than softer X-rays.

This is clearly an important factor in the design of transmitter shielding. In X-ray machines for medical use, metal shields (filters) are used to attenuate selectively so as to maximise the wanted X-ray energies and minimise the unwanted energies.

For radio equipment, where we do not want any of the X-rays, there are complications in shielding high power tubes and the intrinsic electronic tube shields do not normally reduce the doserates to the levels required by the user. Hence there may be a need for further shielding within the transmitter structure to achieve these ends (see Chapter 10).

Both gamma rays and X-rays, being electromagnetic waves, are subject to attenuation in accordance with the the inverse square law and both may travel for a considerable distance. As those who work with ionising radiation will know, X-rays generated by electronic equipment have one valuable characteristic when compared with radioactive sources, namely that the radiation ceases when the power switch is turned off! Consequently, the first line of defence when carrying out surveys and unexpectedly high X-ray levels are encountered, is either to withdraw to a safe distance or to switch off the equipment.

Whilst the level of X-rays commonly encountered by users of transmitting equipment is not to be compared with those met in medical X-ray work or in the nuclear industry, it should be said that very high power electronic tubes can, if experienced unshielded, give rise to very high doserates, as illustrated later in this chapter. However, such tubes normally include some degree of shielding as part of their structure and further shielding is provided in the transmitter.

Ionising radiation units

The SI units for ionising radiation [40] are relatively new to many people. However, changing units normally only affects those already working with ionising radiation, those coming fresh to the subject having no 'unlearning' to do. Appendix 2 lists the old and the new (SI) units together with conversion factors.

The two aspects of basic ionising radiation measurement are:

1 Doserate. The amount of radiation per unit time.
2 Dose. The doserate multiplied by the time duration of the exposure, i.e., the total radiation experienced. It may be applied to a specific exposure or to longer units of time e.g. year, working life.

A simple, though very limited, analogy for those unfamiliar with dose and doserate is a water system where the flow rate (say, litres per hour) represents the doserate and the quantity of water so collected in a given time (litres) represents the dose. The actual calculation of the dose is simply doserate (litres per hour) times the number of hours for which the water flowed. The importance of the dose concept stems from the fact that for ionising radiation there is no accepted safe threshold below which harmful effects do not occur. Thus the effects of ionising radiation are treated as cumulative and all doses are additive as far as potentially harmful effects are concerned.

The basic dose concept has to be qualified in some way to reflect the fact that the harm caused by a given amount of radiation exposure depends on the type of radiation involved. Since we are only considering one type of radiation, it may not be clear why it is necessary to refer to other forms of radiation. The reason that this is relevant is that it necessitates two separate units and it is, therefore, important to the understanding of those units.

The first unit is the gray (Gy) and is used for the absorbed doserate and dose. It is a measure of the exposure rate and dose actually experienced prior to considering the question of the type of radiation involved and the relative harm attributed to a given measure of it. As a doserate the time unit is normally the hour and the unit is the gray per hour (Gyh^{-1}). As an absorbed dose, the unit is the gray (Gy).

Now dose = doserate multiplied by the duration of exposure so that $3\,Gyh^{-1}$ experienced for 2 hours results in a dose of:

$$2 \times 3 = 6\,Gy.$$

In order to determine the equivalence of different types of radiation, a further unit is used, the dose-equivalent rate and dose-equivalent. The unit is the sievert (Sv) and is obtained by multiplying the absorbed doserate or dose by a factor called the relative biological effectiveness (RBE), commonly referred to as the quality factor (QF).

Whilst the value of QF is significant for alpha and neutron radiations, the value for gamma and X-ray radiation is 1. This means that, in practice, when QF = 1, the numerical values in grays and sieverts are the same and only the sievert need be used. The sievert is a large unit and the usual transmitter doserates will be in the order of microsieverts per hour (μSvh^{-1}) with the accumulated annual doses in the order of millisieverts (mSv).

The predecessor of the sievert was the rem and many measuring instruments still carry markings in the old units e.g. rad and rem. These units were those used for absorbed dose (rad) and dose equivalent (rem) before units were rationalised and the new SI units introduced. Sometimes the now obsolete unit the röentgen (R) may be found on scale markings.

It has not been considered necessary to replace this unit in the SI system. However, existing instruments may well last for a long time. In terms of the deposition of energy in tissue, the röentgen is a little smaller than the rad but it is usual to treat the numerical value as equal to the numerical value in rads and rems (see Appendix 2).

In the present context, the millirad, millirem and milliroentgen will be the relevant order of magnitude for normal leakage levels. These can be considered numerically equal. The conversion of millirems, millirads or milliröentgens to microsieverts involves a factor of 10 so that, for example, a commonly encountered specification figure is 0.75 millirem per hour and this becomes 7.5 microsievert per hour.

X-ray hazards

The hazards connected with all ionising radiations are frightening, as well as they might be, considering the possible outcomes. However the user of transmitters and other X-ray producing electronic equipment will not normally encounter very high doserates, assuming that safety instructions are followed. Nevertheless, the full hazards need to be known and understood. These are well documented Nationally and Internationally and need only a brief mention here.

Acute effects are those which result immediately from the dose encountered and stem from the cell damage caused by the power of ionising radiation to eject electrons from the atoms and molecules of human tissue. This cell damage can also result in the production of toxic substances in the cells. Blood cells tend to be affected with a fall in the white cell count. The effects depend on the dose and have been topical in connection with the Russian power station accident at Chernobyl. Obviously at very high doses, the effects can be fatal.

Late effects are those which may occur long after an exposure such as the production of tumours and genetic effects affecting subsequent offspring.

The obvious implication is that ionising radiation must be taken seriously and given adequate attention when designing equipment. At the same time a sense of proportion is required to avoid an excessive fear of it. It needs to be recognised that some uses of ionising radiation are very important for our wellbeing, medical X-rays and scans being good examples of situations where the immediate benefits are likely to be greater than the long term risks. It should be noticed that most existing published material concerning the medical effects at different doses uses the rem as the unit. To relate this to the SI unit, 1 rem = 10 mSv.

X-ray permitted limits

Differing legal provisions apply across different countries and there is no easy way of dealing with this topic. Across the EEC the Euratom Directive applies and is the basis of the Community safety provisions. Many of the provisions are, of course, not relevant here since they deal with radioactive substances.

In this chapter, in order to illustrate the use of limits the UK provisions in the Ionising Radiations Regulations 1985, [41] which meet the requirements of the Euratom Directive, are used. As far as X-rays are concerned, the numeric limits will probably not differ much from those in use in other countries which follow the guidance provided by the International Commission on Radiological Protection (ICRP).

The basis of the applicability of the legal provisions to such things as transmitters is the definition of a 'radiation generator':

"*Radiation generator* means any apparatus in which charged particles are accelerated in a vacuum vessel through a potential difference of more than 5 kilovolts (whether in one or more steps)...". This UK definition concludes by exempting cathode ray tubes and visual display units (VDUs) from some aspects of the regulations, subject to specified doserate limits being met.

The general principle involved is the determination of permitted annual dose limits for various categories of people, e.g. adult males, pregnant females, etc. For the purpose of this chapter, the figures for the adult male will be used. The annual dose limit is 50 mSv. (Although not pursued here it should be noted that the UK body concerned with the provision of advice on ionising and non-ionising radiation, the National Radiological Protection Board (NRPB), is currently seeking to keep this down to 15 mSv per annum average over 5 years and not more than 20 mSv in a single year)[42].

A figure of three tenths of the nominal 50 mSv (15 mSv) is used to define working limits. Dividing this by the notional working hours in a working year (2000) we arrive at 7.5 μSvh^{-1}. A further division by 3 gives 2.5 μSvh^{-1}. With these figures, three categories of working conditions are defined:

Doserate	Implications
>7.5 μSvh⁻¹	The working area must be a 'controlled area'. Dose badges to be worn; regular medical examinations required; individual cumulative exposure records must be kept.
2.5 to 7.5 μSvh⁻¹	Working area to be a 'restricted area'; no other special provisions but all relevant contents of the Statutory Law concerned still apply.
<2.5 μSvh⁻¹	No special requirements for the working area but all relevant contents of the Statutory Law concerned still apply.

The dose badges referred to above are usually thermo-luminescent dosemeters which record the total dose of the wearer and are periodically processed by the issuing body and dose records produced. They generally replace film badges where a piece of film was used for the same purpose. A typical dose badge is illustrated later in this chapter.

It should be said that many employers concerned with the manufacture of high power transmitters use TLD badges voluntarily as a prudent monitoring measure, even though not operating at controlled area doserates. The badges can also be used for certain equipment measurements and this is discussed later in this section.

It can be seen that the basis of the limits shown above provides a choice of working practices with different economic and other consequences for each, for example, the cost of having controlled areas, badge dosemeter costs, and administrative costs relating to record keeping.

Perhaps most important of all is the reluctance of technical personnel to incur unnecessary radiation doses which echoes the general objective of most countries to limit population doses. It is therefore important that new designs have specifically tackled the reduction of X-ray leakage, thus avoiding unnecessary doses to test and operating personnel and expensive remedial design work.

Of course, this structure of permissible X-ray leakage levels does not answer the question most asked – what is the permitted leakage from a transmitter? Rather, it merely fixes the possible everyday working conditions. Fortunately, in practice, transmitter leakage limits are effectively determined by the market place. Users, in general, want as little stray radiation as possible.

This is generally reflected in the specifications of the larger purchasers. In most cases, transmitters are likely to be in the <2.5 μSvh⁻¹ range. Measurement is done at 100 mm distance from the equipment surfaces, except for video display units (VDUs) and cathode ray tubes which, for unexplained reasons, are measured at 50 mm.

The inverse square law applies to X-ray radiation and leaks giving readings of the order of 2.5 μSvh⁻¹ measured at 100 mm from the surface of the transmitter panels will probably be unmeasurable at 200 mm. This means that in many cases to incur even a small dose would probably involve leaning on the transmitter for some considerable time! Of course, the situation could change drastically if a panel is removed and hence the need for strict safety rules and their enforcement.

The transmitter supplier can provide the safety instructions but the user must accept responsibility for their observance. Cases arise from time to time where personnel use their discretion and do something unauthorised in particular circumstances without having the necessary understanding of the risks involved. The authors's past observations include:

1 Personnel breaking a lead-glass window (used for X-ray shielding) and quietly repairing it with a piece of plastic.
2 Temporarily replacing a missing panel with cardboard and overriding door trip switches without checking on the X-ray consequences (the cardboard avoided airflow loss).
3 Cases where small servicing access panels were forgotten and left off when X-rays were present.
4 Lack of servicing screens for 'open door' maintenance adjustments.

At least one extreme case is known where a man removed X-ray shielding and then ran the equipment. In consequence he lost his thyroid gland which was destroyed by the radiation.

X-ray production

As noted previously, X-rays are produced when electrons are accelerated by high voltages within a vacuum and this definition fits the electronic tubes used in high power broadcasting, communications and radar systems. These include triodes, tetrodes, klystrons, magnetrons and travelling wave tubes. Basically, the peak energy of the X-rays will depend on the peak voltage across the tube anode-cathode and the intensity will depend on the square of this voltage. In high power RF circuits this peak voltage will often be greater than the DC supply voltage and can be two or more times the supply voltage. In amplitude modulated systems, the X-ray intensity can, in certain types of system, increase more rapidly than theory would suggest, when the modulation level is increased.

Practical information on X-ray production in X-ray tubes is not difficult to obtain but the same information for electronic tubes is not at all easy to obtain. However, Thomson Tubes Electroniques publish technical information on this subject in respect of their high power pulsed klystron amplifiers. The information supplied is one of the best examples of technical support, in relation to adventitious ionising radiation, which the author has seen from an electronic tube supplier.

The first example of an RF circuit provided by Thomson Tubes Electroniques uses a pulsed klystron operated at high peak power (5.6 MW peak; 5.55 kW mean) and relates to the TH 2066 klystron. Whilst this example may appear to be a very high power case, klystrons currently available can provide effective peak powers of 20 to 30 MW.

This particular test does provide a great deal of useful data. Figure 7.3 shows the result of the tests for X-ray doserates which were carried out on

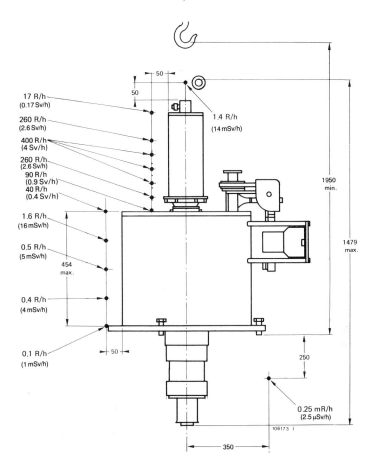

Figure 7.3 *Klystron X-ray test results*
(Courtesy of Thomson Tubes Electroniques, Paris)

the TH 2066 klystron tube. The tube collector has a shield of copper of effective thickness approximately 1.5 cm (upper part of the diagram). This diagram has been left as supplied with units in röentgens per hour but the sievert values have been added in parenthesis (1R = 10 mSv).

With the tube mounted in its electromagnet the highest doserate shown is $400\,Rh^{-1}$ ($4\,Svh^{-1}$) at 50 mm from the tube. Most of the radiation occurs, as shown, in a plane perpendicular to the axis of the collector in the top part of the diagram. From the shielding point of view, it can be seen that the metals in the electromagnet (centre portion of the tube outline) provide substantial attenuation of the much smaller amount of X-ray emission which arises from the body of the tube.

Further shielding will be necessary to reduce the radiation level at the external surfaces of a transmitter or other equipment using this tube under these conditions.

The method used to undertake this type of measurement is to use pre-positioned 'pocket' dosemeters since it is far too hazardous to use personnel. Electronic instruments may be subject to the high level interference in such tests.

Another example provided by Thomson Tubes Electroniques[43] is for a type TV 2002 high power pulsed Klystron with a 240 kV peak beam voltage, a mean output power of 25 kW and a peak output power of 25 MW, considerably more power than the example in Figure 7.3.

The result of the measurements was to establish that the doserate at one metre from the tube was $400\,Rh^{-1}$ ($4\,Svh^{-1}$), much higher than the previous example which gave this level at 50 mm from the tube. It would need considerable attenuation to reduce it to, say, $2.5\,\mu Svh^{-1}$. Whilst the doserate of $400\,Rh^{-1}$ in both examples is very hazardous, it would be difficult to access this level in the first case due to the closeness to a very hot tube. It would, however, be possible to access lesser but still hazardous levels.

In the second case where the distance for the same doserate was one metre, the possibility of accessing this hazardous doserate might be quite high unless effective protection is provided.

It was calculated that a non-filtered X-ray generator with the same beam power would supply about $36\,000\,Rh^{-1}$ ($360\,Svh^{-1}$) at 1 metre away. This gives some idea of the attenuation of the thick-walled collector of the tube which constitutes the filter.

In this example, the applied voltage Vb is 240 kV. It is noted that in this case, as the tube power varies as $V_b^{5/2}$ then the X-ray intensity will vary as $V_b^{7/2}$. Hence, if the value of Vb is increased from 240 kV to 290 kV (an increase of approximately 21%), the X-ray intensity, with the pulse conditions unchanged, will double.

Figure 7.4 from the same literature, shows a source of X-ray radiation from a target bombarded at 240 kV with and without the 1.5 cm copper filter (the thick walled collector of the tube). The copper filter corresponds to that on the tube in the previous example. The Y axis is scaled in percentage spectral density and the X axis is scaled in terms of the tube voltage in kilovolts. It can be seen that the unfiltered spectrum is broad with the peak spectral density at about 125 kV. The copper filter acts as what a radio

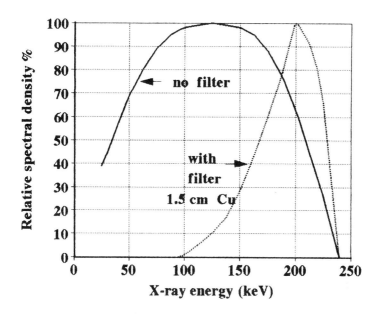

Figure 7.4 *Klystron energy distribution (Courtesy of Thomson Tubes Electroniques, Paris)*

engineer might see as a high pass filter, except that energy rather than frequency is being considered.

Since the X-ray absorption coefficient of copper increases as the wavelength increases (energy reduces) the lower energy part of the spectrum is heavily attenuated, leaving the new energy density curve as one which has a much higher energy at the peak spectral density.

Consequently this is relatively harder to attenuate; that is to say it requires a greater thickness of a given material to attenuate it.

These klystron measurements under high power operating conditions indicate how necessary it is to take X-ray radiation seriously on high power radars and similar equipment. It is clear that very specific safety instructions are needed together with sufficient training for technical personnel to ensure that the warnings are understood.

Figure 7.5 shows a Thomson type TH558 tetrode tube of the type used in high power broadcast transmitters and rated at 650 kW for the MW and LW broadcast bands and 550 kW for the HF band. Data provided in a private communication to the author gives some information on X-ray leakage for pulse duration modulation (PDM) and RF operation and this is listed in Table 7.1. The operating conditions are not given. An additional note stated that the worst doserate outside the transmitter cabinet, measured at 50 mm distance was $0.1\,\mu Sv^{-1}$.

Figure 7.5   *Klystron type TH 558*
(Courtesy of Thomson Tubes Electroniques, Paris)

Table 7.1   *Electronic tube X-ray leakage in pulse duration modulation (PDM)and RF operation – tube type TH558*
   (Courtesy of Thomson Tubes Electroniques, Paris)

Type of operation	CW (μSv^{-1})	Modulation (μSv^{-1})
Close to PDM tube	8	7
Close to RF tube	$\leqslant 0.2$	20

Elsewhere, experiments have shown that with high power HF broadcast transmitters equipped with the type of tube illustrated in Figure 7.5 and using pulsed power efficiency modulation systems intended to provide some economy in the amount of energy consumed, there can also be a rapid increase of X-ray radiation with increase of amplitude modulation depth at the highest depths (e.g. from, say, 80 to 90%) when carrying out measurements with a sinewave test tone.

Figure 7.6 gives an example of this on a high power transmitter. The transmitter already has a fair amount of shielding and the leakage levels were

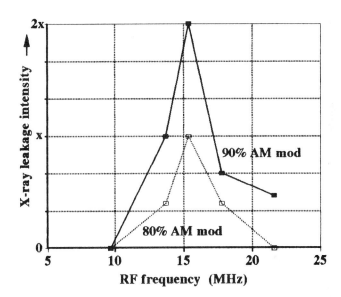

Figure 7.6 *High power HF broadcast transmitter X-ray leakage investigation*

measured with all panels in place and doors shut. The Y axis is not scaled in specific values which, in this case, could be misleadingly low as it applies to a design situation where most of the leakage has already been shielded and only a few remaining locations gave any measurable leakage.

Instead the highest value on the curves has been arranged to be scaled as 2× with × and zero, as the other marked points. By this means, the doubling of doserate can be seen in Figure 7.6. Note that zero here indicates 'unreadable'.

The lines in the graph have no significance as they merely exist to link those test results which are related to each modulation level used. It is important to recognise that X-ray radiation levels can be high where shielding has not already been designed into the transmitter. In such cases the magnitude of the leakage might well be in tens of μSvh^{-1} or more, depending on the structure of the transmitter.

It can also be seen from the same diagram that another factor which is not accountable by normal theoretical considerations arises, namely that the absolute level of X-ray leakage, measured at the same places and under ostensibly the same conditions, shows a dependence on the RF frequency.

This appears to be due to some effect arising from the pulsed operation of the power efficiency circuit and the resulting effective peak voltage across the power tube. An experiment to replace the power output tube by one from a different manufacturer merely displaced the RF frequency at which the X-ray peaked by a small amount. A paper by Hunter[44] attempts an analysis of the effect of pulsed power efficiency systems and highlights the fact that bursts of X-rays appear to be generated on the rise and fall edges of the modulator pulses.

The implication of the foregoing is that when surveying such amplitude-modulated transmitters, especially very high power ones, the tests must include the higher modulation depths and a range of RF test frequencies such that any peaking of the X-ray will be spotted. The author did not find any effect attributable to the audio modulating signal frequency and 1 kHz is a quite usual test frequency.

Testing at the highest modulation levels on very high power transmitters can introduce problems resulting from the fact that such transmitters may have absolute time duration limits for operation at 100% modulation, since the normal average modulation level is much less. The main problem is to survey the surfaces of large cabinet structures in the allowed time limit. The need to test at a number of frequencies does not usually impose any extra burden since RF leakage tests require this and will usually be carried out in conjunction with the X-ray tests.

Measuring equipment

The measuring equipment used for X-ray measurement may be unfamiliar to those who have not had to involve themselves in ionising radiation. There is a considerable variety of types of instrument available and the basic operation of these is covered in the following paragraphs.

Ionisation chamber instruments

The basis of the measurement of X-rays in this type of instrument involves the determination of the amount of ionisation caused in a gas-filled chamber.

Ionisation involves the production of ion pairs consisting of a negative ion (electron) and a positive ion. The chamber may use air or other gases, the air chambers usually being vented to the atmosphere.

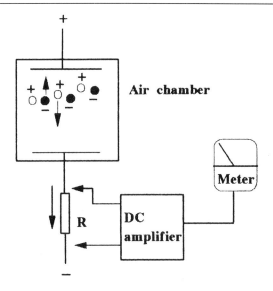

Figure 7.7 *Ionisation chamber X-ray measuring instrument – basic concept*

Figure 7.7 illustrates the general principle of operation. The chamber has two conductive plates, one being connected to the high voltage positive DC supply terminal and the other to the negative supply terminal. The negative (return) connection is via a high resistance R. The chamber has a thin window, not depicted in the diagram, which minimizes the attenuation of X-rays at the low energy end of the spectrum.

When the chamber is exposed to an X-ray source, ionisation takes place as indicated in Figure 7.7. The positive voltage on the upper plate will attract the negative ions (electrons) and the positive ions will be attracted to the lower plate.

As a result a very small current which will be proportional to the number of ion pairs produced and thus the X-ray doserate, will flow through resistance R. As a result, it will develop a corresponding voltage across it. As the currents are very small, amplification is required to provide a current sufficient to operate a moving coil meter. The current flow arrow in Figure 7.7 indicates conventional current flow (positive to negative).

It is necessary to ensure that the X-rays fill the chamber window otherwise some error may result. However, in the author's experience using several instruments of differing types on a variety of transmitters, very close agreement was found on each occasion even though the instruments had different chamber aperture sizes.

The meter can be scaled in doserate. With the newer instruments, scaling in sieverts is usually available. Sometimes the chamber is made of plastic, filled with a gas and sealed. This arrangement can have a good low energy response down to about 6 to 8 keV. It can, however, suffer from the loss of the gas due to permeation through the plastic and may need periodic recharging with gas.

Purchasing a spare at the same time as the instrument is purchased is not necessarily a help since this may suffer the same gas loss. If a spare is needed, it should be purchased at about half way through the 'charge life' of the original chamber.

It is generally agreed that the ionising chamber instrument should always be used for transmitter X-ray measurements. A type appropriate to the work must be chosen.

The Geiger-Müller counter

The Geiger-Müller (GM) tube is a much more sensitive device than the ionisation chamber. It is basically a gas-filled tube operating at a voltage below that which would sustain a continuous gas discharge current. Figure 7.8 shows a much simplified representation of a GM tube where the positive

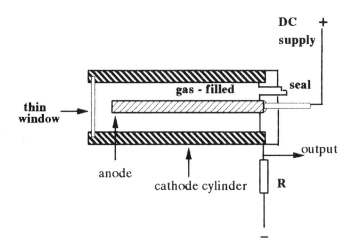

Figure 7.8 *G-M tube X-ray counter tube*
(Courtesy Philips Electronic Tubes)

plate is the central conductor and the negative plate is the cylindrical cathode. The pulse voltage is developed across the resistor R and coupled to a pulse threshold detector in the instrument circuitry and then into a counter. The tube is filled with a suitable gas, argon or krypton typically being used for tubes intended for X-ray work.

An ionising particle or X-ray photon will cause the production of an ion-pair as before in the ion chamber but in this case the electrons so produced will produce further ion pairs due to the extra acceleration provided by the higher supply voltage, resulting in an avalanche of current. This is known as gas amplification and results in a large pulse of current.

It is important to ensure that only one pulse is counted for each initiating photon, so a small amount of another gas, usually one of the halogen family, is added. This is known as a quenching gas and prevents possible spurious discharges which can result from secondary electron production in the tube.

The effect is to render the tube insensitive for a short period between pulses and this is known as the 'dead time' after a pulse during which a successive pulse will not be counted. In a selection guide from Philips Electronic Tubes[45] the magnitude of the 'dead time' ranges from 11 to 230 μs according to the type and function of the tubes. Dead time needs to be taken into account when the manufacturer calibrates the instruments.

The energy response of GM tubes is not very flat and in order to flatten it, some tubes are fitted with a filtration cylinder consisting of an arrangement of several types of metal.

Specially designed tubes such as those listed by the manufacturer mentioned above, offer X-ray low energy operation down to about 2.5 keV.

When using a Geiger-Müller instrument on amplitude-modulated transmitters with high depths of modulation, the guidance given in the instrument handbook should be followed regarding the modulation test tone frequency to avoid the instrument registering the modulation frequency rather than the X-ray radiation.

One instrument illustrated later in this chapter uses a G-M tube. G-M tube instruments are generally much cheaper than chamber instruments to manufacture, but tend to have a poorer energy response at low energies.

Scintillation counters

Another type of instrument is the scintillation counter (Figure 7.9). This typically uses a sodium iodide crystal doped with tellurium or thallium. When the crystal is subjected to X-rays it gives off flashes of light which are amplified and detected by a photomultiplier tube. The dead time on this type

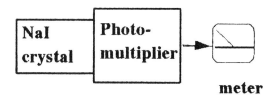

Figure 7.9 *Scintillation counter instrument concept*

of instrument can be around 4 µs. This type of detector is very sensitive but some types lack a good low energy response.

Both the G-M tube counter instrument and the scintillation counter, when suitably chosen for the purpose, are particularly good for the initial part of a survey where they act as 'sniffers' (search instruments to locate leakage beams).

Other instruments and devices

The quartz fibre electrometer

Earlier in this chapter a reference was made to pocket dosemeters. The type referred to is one which has been long established, the quartz fibre electrometer. This is technically also an ionisation chamber instrument but made in a size and shape similar to a fat fountain pen. It is a cylindrical device typically about 10 cm long and about 1.3 cm diameter. It consists of an ionisation chamber with a direct reading quartz fibre electrometer on the lines of a gold-leaf electroscope. In this case, the quartz fibre is repelled by a fixed electrode. The fibre can be viewed through a lens against a scale. The device has to be charged with a voltage before use.

When subjected to ionising radiation the device will discharge, the discharge being proportional to the doserate of the radiation received. The dose is read off the scale viewed through the lens.

The purpose of the device is to give an immediate reading of dose so that when worn by personnel, information about their exposure is quickly available. It can, as in the earlier reference in this chapter, be used in an experimental situation to record dose without having to expose personnel.

Because it has no electronics circuitry it is immune from the sort of problems which might occur with high levels of RF present if an electronic instrument was used. However, because of its nature it is susceptible to errors if roughly handled or if dropped. It is often worn by personnel as a means of getting a quick indication of radiation present, as contrasted with the thermo-luminescent dosemeter (TLD), which has to be sent for processing before the information is available. It is usually backed up by a TLD.

Film and thermo-luminescent personal dosemeters

For many years a 'badge' type dosemeter using photographic film was used to record radiation doses for people working with X-ray, gamma and other forms of radiation.

The badge was processed at intervals and the dose recorded. The dose information was obtained by comparing the darkening of the relevant part of the film for the radiation concerned with a sample of the film from which the badge was made.

In the UK and some other countries, the badge now contains a quantity of a substance which exhibits thermo-luminescence such as lithium fluoride.

This means that the ionising radiation dose recorded by this substance can be converted to a light emission proportional to that dose by heating the substance in conjunction with a suitable measuring system. The system operated by the UK NRPB uses a badge which covers 0.02 to 2 MeV for photon radiation with an uncertainty of about ±25% and a measurement threshold of about 0.1 mSv. It also covers beta radiation, but this has no relevance in this chapter. A typical badge dosemeter is shown in Figure 7.10. The badge is officially called a thermo-luminescent dosemeter (TLD) though, out of force of habit, people still call them film badges.

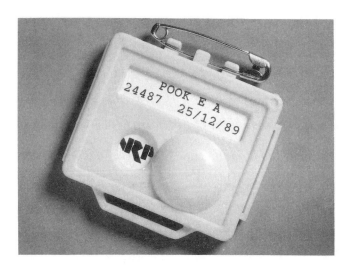

Figure 7.10   *Thermo-luminescent dosemeter (TLD)*
(Courtesy National Radiological Protection Board)

An additional and important use of the badge is for equipment testing where the badge can be attached to the equipment being investigated and the equipment run for a specific time. The resulting dose reported by the dosemeter processing organisation is divided by the equipment run time, in hours, to give an average X-ray doserate. Most badge processing services provide for rapid assessment of badges reserved for investigations and report by telephone.

Example:
Dose recorded = 0.5 mSv; time equipment run = 10 hours;
   Hence the doserate = dose/time = 0.5 mSv/10 = 50 μSvh^{-1}.
   It is obviously important to ensure the accuracy of the run time record otherwise the calculation will result in an average doserate which is too high if the time is understated and too low if the time is overstated.

Note that the quartz fibre electrometer can also be used for such experimental measurements, as illustrated in the klystron X-ray measurements discussed earlier. It has the advantage of being directly readable.

Practical instruments

Practical instruments are more sophisticated than has been illustrated and have range and function switching facilities giving a wide range of doserate measurement. Meter zero adjustment, and battery voltage checking are also provided. Some also offer dose measurement by integrating the doserate over a specified time. In addition to X-ray and gamma ray measurement, they generally offer some alpha and beta radiation measurement facilities. Whilst the extra measurement facilities are not directly relevant to the assessment of X-rays in transmitters, they might influence the choice of instrument if the user has other ionising radiation measurement commitments.

Most people involved in the measurement of X-ray radiation from radio transmitters recognise the value of two different types of equipment for the task. The first type, referred to earlier as a 'sniffer', is a sensitive instrument, often not calibrated, the sole function of which is to search for leaks so that they can be measured later. The main requirement is sensitivity over the appropriate energy range and portability. An audible indication proportional to doserate in addition to the meter indication is very useful since, in many large surveys, it is quicker to get a first indication aurally rather than alternately having to look at the equipment under test and the instrument meter. It should be able to detect narrow X-ray beams which might be missed by the normal measuring equipment. It will usually take the form of a Geiger-Müller counter or a scintillation counter.

The second requirement is for an ionisation chamber type instrument with a suitable sensitivity with which to perform the measurement of leakages found by the sniffer instrument. There is a very wide range of choice of instruments capable of meeting these two instrument requirements in the world market since the business of monitoring ionising radiations has been established for many years. For those unfamiliar with such instruments, careful scrutiny of suppliers' data sheets plus practical trials are recommended. Because of the nature of radio transmitters and their supply voltages, there is a need to ensure that the low energy response is adequate. Many instruments do not have an acceptable energy response down to the 10 or 15 keV likely to be needed.

This is because they have been designed for higher energies such as those experienced from radioactive substances.

Another special requirement for radio transmitter work is that the readings obtained on the X-ray doserate meter shall not be affected by RF fields. This does not mean that it has to be specially designed since some existing instruments do exhibit a considerable rejection of RF by their nature.

Some instruments specifically claim rejection of RF but, in a few cases, the energy range offered may not be adequate for the work. This can be due to a worsening of the low energy end of the energy response curve, due to the added shielding. There has recently been an improvement in the availability of 'RF proof' instruments which have retained their low energy response.

RF rejection should be judged on the basis specified by the supplier and related to the proposed use. It is unreasonable to expect the rejection of the enormous RF levels which might be found inside a high power transmitter when panels have been removed!

The instruments illustrated in this chapter are typical ones and many, but not all, of them have been used by the author. However any reference here to an instrument is solely intended to highlight features and does not imply any recommendation. Equally, there is no adverse implication in the absence of any instrument manufacturer or instrument.

The actual choice of instruments should be made against the specific requirements of the work and the nature of the radiation to be measured.

Energy responses illustrated here together with other data, should be treated as illustrative. Where energy responses have been illustrated, the energy range covered is only part of that specified by the manufacturer (to 100 keV) since this is of most interest for transmitters. The supplier's data sheets should be consulted for the latest definitive information, which may be subject to change from time to time.

The number of types of instrument per supplier illustrated here is obviously limited, but most suppliers have a range of instruments available covering two or more of the categories mentioned here.

Search instruments

Geiger Müller counter

Figure 7.11 shows the Minimonitor type X, a relatively low cost instrument specifically designed as an X-ray leakage search instrument. The instrument is light and portable. It is not calibrated in doserate but has a 'counts per second' scaling on the meter together with an audible indication of count rate. It is intended to be used as a 'sniffer' and used for comparisons, not measurement.

In carrying out initial surveys over large areas of equipment surface it is often impracticable to keep watching an instrument meter and it is usual to listen to the audible note. For noisy environments, a headphone socket and headphones are available as an optional extra.

The instrument uses a G-M tube as the basis of operation and the window area of the probe is 2.25 cm². The gamma response is given as typically 2 counts per second per μSvh^{-1}.

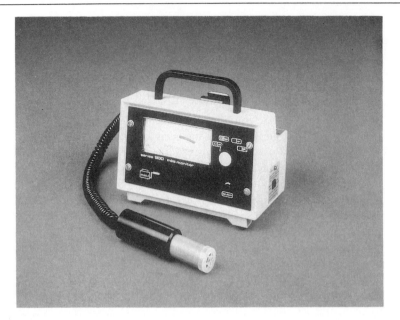

Figure 7.11   *Minimonitor type radiation meter designed for finding X-ray and gamma radiation leaks*
   (Courtesy Mini-Instruments Ltd)

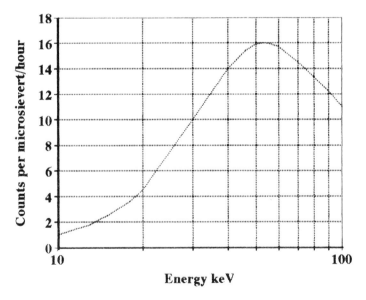

Figure 7.12   *Minimonitor type doserate – energy response*
   (Courtesy Mini-Instruments Ltd)

Figure 7.12 illustrates the typical response versus energy for the range up to 100 keV. The published response actually goes to 1 MeV. The 'Y' axis marking is in terms of counts per μSvh^{-1} where it can be seen that above about 15 keV the Figure of 2 counts per μSvh^{-1} is conservative. The operating conditions for Figure 7.12 were with the probe end-on to the radiation and the plastic cap fitted.

An alarm is fitted (analogous to the alarms on some RF radiation meters) which is set to full scale. It should be noted that ionising radiation instruments are normally, as a general policy, made to read over full scale when overloaded as this one does and the alarm is therefore a warning against encountering an unexpected high level. Dry batteries or rechargeable cells are options. The use of mains-operated equipment is undesirable because of the likely pick-up of RF in the mains lead.

Scintillation counters

Scintillation counter probes are available from the Mini-Instruments company and from the Victoreen company. Both have versions suitable for low energy X-ray and gamma measurement.

Ionising chamber measuring instruments

Figure 7.13 illustrates the Victoreen model 471 X-ray survey meter. It does, of course, cover some alpha, beta and gamma radiations as well, but the

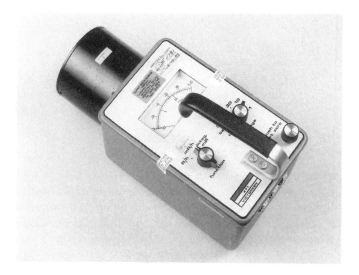

Figure 7.13 *Victoreen doserate meter type 471 (Courtesy Victoreen Inc.)*

shows that at the low energy end there is a penalty in that below 30 keV the response falls off appreciably, the specification for energy response being within 10% from 30 keV to 2 MeV.

However, in contrast to the last-mentioned example, another type claiming shielding against RF is the Victoreen model 440RF/D which is designed for use in measuring television sets and offers a good energy response within 10% from 12.5 to 42 keV, increasing to 40% at 100 keV. It also has the same most sensitive measurement range (0 to 10 μSvh^{-1}) as other instruments mentioned here.

Figure 7.15 illustrates the Eberline RO10 ionisation chamber instrument. This has an appreciably flat energy response from about 12 keV to 1 MeV. The most sensitive range is 0 to 10 μSvh^{-1} and the scale is marked in microsieverts per hour.

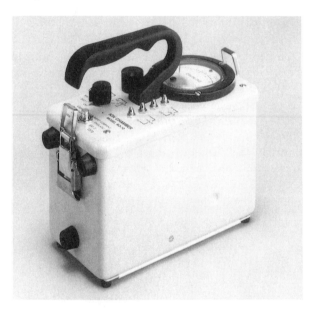

Figure 7.15   *Eberline type RO10 doserate (Courtesy Eberline Company)*

The design is ergonomically interesting since the window to the chamber is, when the instrument is standing as shown in Figure 7.15, on the underside of the instrument. This means that when the instrument is held with the chamber window facing the source of radiation, the meter is vertical and very easy to read. The energy response to 100 keV is shown in Figure 7.16.

There is a sliding Beta shield over the chamber window. This also acts as a protector for the thin chamber window when the instrument is put down on a bench which has objects on it, such as nuts and bolts!

interest here is the X-ray measurement. The instrument uses an air-vented chamber, forming a front projection on the instrument. It is provided with a Mylar end window. A plastic cover (beta cap) also provides protection for the Mylar when not in use. Judging from the published energy response, this cover is best off for the energy levels up to about 60 keV and on for the higher energies.

The most sensitive doserate range is 0 to 10 μSvh^{-1}, a very useful range for transmitter work. It is also available marked in sieverts, which avoids conversions for those working in SI units.

It also has an integrating function which can record dose by measuring doserate over a time period. The instrument is portable and battery operated.

In the data sheet, the actual energy response curve is shown up to 1 MeV but only the part most relevant to radio transmitters (10 to 100 keV) is shown here in Figure 7.14. The actual specification for the lower energy end

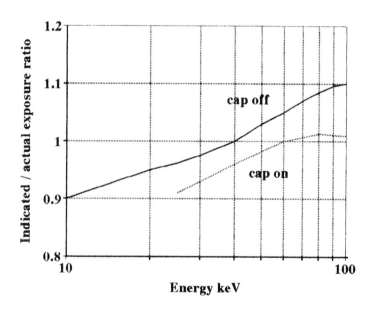

Figure 7.14 *Victoreen meter type 471 energy response (Courtesy Victoreen Inc.)*

of the range is within 10% from 6 keV to 300 keV with the cap off. The full specification (within 10%) continues, with the cap on, to 2 MeV.

There is another version of this instrument, the model 471RF, which, in appearance, is the same as the type 471. However, the model 471RF claims to be specially shielded against RF energy. The published energy response

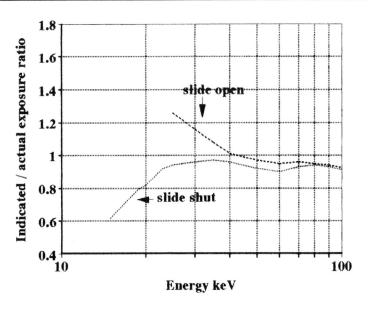

Figure 7.16   *Eberline RO10 energy response (Courtesy Eberline Company)*

Choosing instruments

For the purposes of this chapter it is assumed that the application is for transmitter or RF machine X-ray measurements only. Those having other ionising radiation responsibilities may wish to include in their choice criteria other measurement capabilities, such as beta and higher energy gamma measurements.

Checklist for ionisation chamber measurement instruments

Doserate and dose ranges available

Most people are likely to need to resolve down to $1\,\mu Svh^{-1}$ so that a full scale deflection of $10\,\mu Svh^{-1}$ or less is desirable. The choice of other doserate ranges required depends on the proposed usage. Ranges covering hundreds of millisieverts per hour or greater are unlikely to be needed.

If dose measurement is also required i.e. integrating doserate over a period of time, then this will be a choice factor. The author has rarely found this to be of great value in practice since it takes up the time of the person undertaking the survey and can be covered in other ways. However this may not always be the case, particularly with very low level measurement which can be done by doserate integration.

Scaling of the instrument

Factors of interest are the units used for scaling and the readability of the meter scales. Some suppliers offer scaling in sieverts on request. Such scaling is not essential but does remove a possible source of errors in converting from other units. Rads, rems and röentgens are obsolete units and best avoided if a choice is available. (The fact that some current product test specifications use old units should not inhibit the purchase of instruments with a sievert scaling.)

Energy response

A reasonably flat energy response from 100 keV to 10 or 15 keV is desirable. The upper limit will depend on the high voltage supplies to the transmitters concerned. The need to have a good low energy response stems from the fact that 5 kV and above is usually invoked in the definition of radiation generators and that a lot of transmitters have their high voltage supplies in the 5 to 10 kV range.

Physical characteristics

X-ray surveys can take quite a time on large installations and the weight of the instrument, ease of handling and reading it and similar ergonomic factors become quite important. Note that when measuring X-ray leakage at a specified distance from a panel or other surface, the distance is measured from the centre of the chamber and not from the end of the instrument nearest to the leakage being measured. Instruments have some form of mark or other indication to show the centre of the chamber. This is illustrated in Figure 7.17.

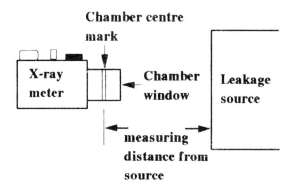

Figure 7.17 *Ionising chamber instrument reference datum for measurements*

Reliability

This is, as with any other instrument, a very important factor but one which cannot be determined from data sheets. The cost of ownership of both RF and X-ray measuring instruments can be very high if they prove to be unreliable since each repair, except the most trivial, will generate a requirement for a re-calibration, the cost of which can be comparable with the repair cost. Indeed, if the instrument is prone to failure, the inevitable repair-recalibrate sequences may result in the equipment rarely being available and a further instrument being needed. The only practical advice on this topic is for the prospective purchaser to talk to other people known to use the instrument in question.

RF protection

It is important that the reading of the instrument should not be corrupted by the effect of any RF present. A few instruments make claims for RF rejection, although some types prove satisfactory without any claims being made. In practice, it is normally easy to determine whether RF is affecting an instrument by judicious use of aluminium cooking foil, as described in Chapter 9.

Energy assessment

The measurement of the effective X-ray energy in a detailed sense would imply the evaluation of all the elements in the energy spectrum and is impractical and unnecessary. There are some useful basic concepts in use in this field which provide a way of evaluating the effective radiation energy. The reason for wishing to find the effective energy of the beam is that if shielding is needed it is this value which is needed to determine what thickness of metal to use.

The basic technique consists of using standard metal test plates of thicknesses specified in the IEC standard IEC 762[57] and inserting these in the X-ray beam in turn until one of them causes the X-ray doserate meter reading to fall to exactly half of the reading obtained without any plate in use. The meter is of course kept in the same place throughout and the transmitter operating conditions kept constant.

The effective energy is read by looking up the metal thickness of the relevant test plate in a table in the standard which lists energy versus metal type and thickness.

Since the subject of X-ray energy measurement and shielding is described in more detail in Chapter 10, which includes reproduction of the IEC762 table, the matter will not be discussed further here.

8
Planning surveys and measurements

Introduction

Surveys can be classified into three broad headings according to their nature and purpose. These are:

1 RF leakage tests for unintended radiation from transmitter cabinets, antenna exchanges, loads and other items connected to RF transmitters; leakage from RF process machines, medical apparatus, microwave ovens and other sources of RF energy. With the present interest in radiation from such things as video display units, which might not be seen by most people as RF sources, leakage tests may extend to a wider range of sources than has been the case in the past.

2 X-ray leakage tests for any transmitters and other RF sources which use voltages higher than 5 kV to operate electronic tubes.

3 RF exposure tests to establish the potential exposure of people to fields from antennas or other systems which are intended to radiate RF energy.

Surveys may involve some or all of these activities. Microwave oven testing is an example of simple leakage testing. X-ray testing will mainly arise on high voltage transmitting equipment but can also arise with video display units and similar devices associated with transmitting systems or other RF sources.

For RF exposure surveys there are two basic approaches to exposure testing, differing only in objective and not in technique. These are:

1 The characterisation of an antenna and transmitter as a portable entity which can be deposited at an unknown site and so located that clearly defined safety limits can be observed. It follows that this is only fully achieved when there is no likelihood that the equipment will be deployed amongst buildings or other structures which could invalidate the original radiation safety data. In order to do such characterisations, it is necessary

to use a flat site free of any structures and buildings which might seriously affect the results, so that the survey data is as 'site independent' as possible.

2 Measurements made at a particular place and time to determine safety provisions applicable only to that place. This type of survey takes into account the real environment – buildings, other antennas, people and their work patterns and is the most common type of exposure survey.

In order to characterise a system, the ideal test site is free space but the practical site will generally be a flat unobstructed area sufficient in size to allow operation under the worst case radiation conditions. A great deal of measurement will be necessary and suitable safety contingencies need to be added in determining the final computed safety zone.

It can be seen that this approach allows both a system producer to specify some definitive safety provisions such as a prohibited volume of space in the equipment handbook and a user to do the same in a safety management manual. This approach does not, of course, eliminate the need for specific safety measurements to confirm the safety aspects since these are required by any safety management system to meet the legal duty of care. What it can do is to diminish the total amount of safety survey work, compared with that which would be needed when starting from scratch.

The same approach can be applied to some other systems such as small communications antennas with short hazard ranges and intended for siting in relatively uninhabited areas such as roof tops. Here an adequate allowance for additive reflections can be made and the hazard area fenced off.

The second case is the much more common one from the point of view of the equipment user, where each site for antenna and transmitter systems has unique characteristics in respect of topography, buildings, structures and personnel movements. Each site needs to be treated as a unique case. The supplier's equipment handbook may only be able to convey general warnings such as details of the maximum power involved and similar basic data.

This chapter is arranged so that general preparation and planning aspects common to all the three headings above are dealt with first, and then those planning aspects specific to a particular type of survey are discussed. Carrying out surveys and measurements is then dealt with in the following chapter.

The selection and preparation of measuring instruments

RF radiation meters and X-ray doserate meters should be chosen to be suitable for the range of work to be undertaken, as discussed in Chapters 6 and 7.

In the case of RF radiation meters, this means coverage of the full frequency range needed, facilities for measuring the appropriate quantities, power density, electric field and magnetic fields in accordance with the relevant standard. It also means taking care when undertaking the measurement of pulsed transmissions that the 'peak' pulse power will not burn out the chosen instrument.

Additional equipment such as spectrum analysers and wideband measuring receivers should, if required for the survey, be chosen according to application e.g. identification of signals which interfere with the survey, etc.

Wherever possible, duplication of the radiation survey meters is recommended, particularly when away from base. Anyone who has travelled several hundred miles with a single instrument which has became unserviceable in transit will appreciate both the economic consequences and the embarrassment resulting! It will not usually be necessary to duplicate expensive supplementary equipment.

Dry batteries constitute one of life's regular problems. Spares often turn out to have passed their prime when suddenly called into service. Rechargeable batteries appear to be more helpful providing that the charger or charger lead is not forgotten. However they do need to be watched for loss of capacity and replaced when they no longer hold their charge.

Some sort of check source which generates a small RF field is very desirable to check functional performance and ensure that probe cables are not broken or intermittent. These devices are available commercially from the instrument suppliers or may be improvised locally.

For HF and VHF frequencies a small oscillator and coil will suffice and will be a little more portable. Technology problems can be minimised by using the lowest frequency the instrument covers as the check frequency.

X-ray ionising chamber instruments need their desiccators checking frequently as moisture can interfere with the instrument operation due to leakage currents occurring where the impedance is required to be high. It is also important to ensure that the chamber window is not damaged.

There is no easy way of providing a check source for X-ray instruments without carrying a radioactive source, and this may be subject to specific regulations regarding handling, transportation and safe custody unless the activity is low enough to secure exemption from some or all of these aspects. This clearly depends on national legal provisions. Certain radiation doserate meters of German origin incorporate a small source for setting up, in which case no other check is needed.

Apart from the main instruments dealt with in Chapters 6 and 7 there are a few ancillary items – tools, materials, and similar items which can be necessary on surveys. These include:

- Tape measures.
- A modern electronic (ultrasonic) measuring 'tape' can be useful indoors

where walls and cabinet structures can be used as reflection media when measuring equipment in order to make drawings for a survey report. Outside use is normally impractical and, in the author's experience, electronic measuring 'tapes' may not survive acquaintance with a high power radar!

- An optical range finder can be useful for outside work if of a robust type.
- A roll of aluminium cooking foil is indispensable for experimental shielding, wrapping round the RF probe when zeroing the instrument if this is proving difficult and for checking whether RF is interfering with X-ray instruments.
- Self adhesive aluminium strip is also widely available and is useful for temporary shielding around waveguide flanges. For X-ray purposes, self adhesive lead strip is also available.
- A roll of adhesive tape is essential to put leakage markers on equipment. The tape must be specified as a tape which will not damage paint finishes on equipment. Many commonly used domestic tapes are likely to remove the paint finishes, something which is not appreciated by the equipment owners!
- For outside work, objects are needed to act as markers at measured positions round the antenna system which is being surveyed. Plastic 'traffic cones' are particularly good for this as they stack easily for transportation.
- For pulsed transmission, a cheap and excellent detector of the pulse repetition frequency (p.r.f.) is any small medium or long wave personal portable radio which will pick up the p.r.f. It is also probably the simplest checking device to verify that radar sector 'blanking' is properly set (blanking is described later).

Practical use of RF radiation instruments

The handling and manipulation of instruments

With instruments which have a separate probe and instrument unit, the probe should normally be held well forward in front of the body and the meter unit held by the side where it can be read. The object is always to minimise perturbation of the field. Isotropic probes, which would be the normal type for most work, should be pointed at the source and rotated on their axis through 360° to peak the isotropic response and thereby reduce the effect of isotropicity error.

For instruments having a rigid probe plugging in to the meter unit so that they are effectively one unit, the probe should be pointed as above but the task of rotating the probe on its axis is more difficult as the meter will disappear from sight in the course of rotating it through 360°. This type can also give similar problems when measuring leakage in awkward situations. On the credit side, they do offer single-handed operation in a situation where one does not seem to have enough hands.

Some instruments with probes which fix rigidly offer the option of using a flexible connector, in which case manipulation is as in the previous paragraph and operation becomes two-handed!

Regular zero setting of the instrument is required for most types and this involves taking it out of the field or using a metal shield or the ever useful cooking foil to cover the probe. If this is done without adequate shielding, then any stray field present may be backed off so that when, later, a measurement is made at a place where that field is low or not measurable, a reverse meter reading may be experienced. Consequently, it may also necessitate repeating much of the work. There are some instruments available where the maker specifies that zero setting can be done in a field. The maker's handbook should be consulted.

Where instruments have a 'maximum hold' facility this needs to be used with the correct time constant selection (usually the fast setting) as instructed in the maker's handbook. The facility is very useful for the initial appraisal of potentially hazardous fields such as a stationary microwave beam. If the facility is left switched on when it has been finished with, it can cause considerable confusion!

The build-up of electrostatic charges on instruments can have a nuisance value since it gives transient readings which disappear gradually and usually result in an unsuccessful repeat search for the illusive reading. The great danger, with RF instruments, (and ionising radiation instruments for that matter, as they are also susceptible) is that it may be assumed that every sudden reading is due to static, and consequently leaks may be missed. Patience in allowing charges to discharge is what is needed.

Uncertainty of measurement

The term 'error' is sometimes used when it is not justified. This term can only be used definitively for systematic errors and only then where the magnitude and sign is known. Random errors cannot be known and, therefore, corrected. In practice, the limited knowledge of systematic errors over the range of environmental conditions in which an instrument is used coupled with random errors results in the assessment of an 'uncertainty of measurement' to be attributed to an instrument.

Amongst those who do practical measurements there seems to be a broad agreement that the practical uncertainty of measurement of electromagnetic fields is around ±30% and ±40%, where the variation of the calibration correction factor from unity is small or has been applied as a correction.

The use of correction factors from the instrument calibration certificate or from information marked on the probe is important when the correction is significant. Otherwise, a large uncertainty of measurement might have to be attributed. When the measured values are near to the permitted limits it is important to allow for the uncertainty of measurement of the instrument otherwise it is possible that the permitted limit might be exceeded. This is a basic measuring concept and is not peculiar to RF field measurement.

It will usually be better to apply all corrections after the actual meter readings have been recorded so that any arithmetic errors can be detected and corrected by reference to the recorded figures and the known calibration factors. On those instruments which have a control which can apply calibration corrections by manual application, there is the option to correct at the time or to set the correction control to unity and apply the calibration corrections later.

Where duplicated instruments are used on surveys, it cannot be expected that they will give identical readings since two instruments calibrated with an uncertainty of $\pm x\%$ can, theoretically, differ by up to $2\times$ after the application of any documented correction factor.

If two instruments have markedly different calibration factors it would only be reasonable to expect them to give different readings for particular measurements. This may cause concern if the calibration factor data has not been scrutinised beforehand, but when the relevant factors are applied it will usually be found that the readings begin to converge and be as close as is reasonable, taking into account the calibration uncertainties.

Sometimes the calibration is suspected when the real cause of an apparent discrepancy is something different. This can be the case in a location where there is a considerable change in field with a small change in distance and measurements with two instruments have not been made with sufficient care so that they were, effectively, taken in different places.

When out on surveys, suspect instruments can be best checked by comparison with a known good one using a test source and taking care to position each probe identically, if possible using an improvised locating jig such as a slotted piece of wood or something similar.

A further problem can arise in the recording of measurements. It is generally appreciated that the dynamic range of RF radiation measuring instruments aligns with RF protection guides so that 'unreadable' cannot always be interpreted as meaning that there is no leakage. In fact there may be enough leakage to cause EMC problems although well below safety levels for human exposure.

It is therefore important that, unless there is no possibility of misunderstanding, it is better and more accurate to record as 'not greater than y' where 'y' is the lowest reading that can be resolved on the instrument. Where there is a general RF ambient level in the area where measurements are being made, that level will limit the lowest resolvable reading anywhere.

For example, if the general ambient is $1\,Wm^{-2}$ then the lowest recorded value possible will be 'not greater than $1\,Wm^{-2}$' even if there is, in fact, nothing being produced at the point being tested. Consequently, where the ambient background is sufficient to be an embarrassment, it will be necessary to attempt to secure a reduction by switching off unnecessary equipment and possibly by negotiating an 'out of service' maintenance time slot for the troublesome sources. With modern transmitter installations the ambient levels inside the station will rarely be a problem, but with MF and HF stations using open wire feeders internally there may be an appreciable nuisance.

Where there is an EMC interest, it is likely that this would need to be investigated with a measuring receiver or other sufficiently sensitive instrument, because of the limited sensitivity of RF radiation meters. For example, with a particular instrument with a lowest range of 0 to $20\,Wm^{-2}$ (0 to $2\,mWcm^{-2}$), the maker specifies the measurement range as starting at $1\,Wm^{-2}$ ($0.1\,mWcm^{-2}$).

Now, from the human safety point of view this level of RF field is not a problem. However, as a potential source of interference, $1\,Wm^{-2}$ corresponds, on a plane wave basis, to an electric field of just over $19\,Vm^{-1}$ which, when considering sensitive equipment, may still give rise to problems, particularly with receivers. Again, $19\,Vm^{-1}$, according to British standard BS6656 (see Chapter 5) can, on a CW basis, be hazardous to some flammable vapours up to about $300\,MHz$.

Some modern instruments can offer an order better in sensitivity, but often this is, in the nature of things, at the expense of the CW and pulse overload absolute limits.

Avoiding sensor burn-out in RF radiation probes

The point has been made earlier that both diode and thermocouple sensors in probes can easily be destroyed by excessive fields regardless of whether the instrument is in use or not. The specifications for probes include maximum safe ratings for CW overload and peak power density. Where a probe is intended for pulse transmission work, the peak power density rating is very important.

Two examples taken at random from a manufacturer's catalogue are shown in Table 8.1. They were chosen to give different safety factors but are typical probe ratings. The parameters given are in this example are:

CW maximum overload rating
Peak pulse power density rating

The table also gives the greatest measurement range meter full scale deflection (f.s.d.) value. For the purposes of this paragraph, the units used

Table 8.1   *RF instrument peak power limitations*

Probe	Greatest range power density (meter f.s.d.)	CW overload power density (mWcm^{-2})	Peak power density (Wcm^{-2})*
A	20 mWcm^{-2}	60	20
B	100 mWcm^{-2}	1000	300

*Note unit used.

are those used in the catalogue, since they involve hybrid units and most people will find it easiest to relate these to the mWcm^{-2} unit used for f.s.d. values.

The CW overload level is 3 and 10 times the f.s.d., respectively. The real problem arises with pulse transmission measurements.

Taking probe A as an example, assume a calibration correction factor of 0.95 at the measurement frequency. On the highest range of the instrument (f.s.d. = 20 mWcm^{-2}) the mean value (meter reading) corresponding to the peak power rating for a duty factor of 0.001 will be:

$$S_{max} \text{ (mWcm}^{-2}) = (0.001 \times 20\,000)/0.95$$

Note that the second number (20 000) is 20 Wcm^{-2} expressed in milliwatts per square centimetre; S_{max} is the maximum permitted meter reading (mWcm^{-2}).

Hence $S_{max} = 21$ mWcm^{-2} which exceeds the f.s.d.

The instrument should be kept to 20 mWcm^{-2} maximum, i.e., not allowed to go overscale.

Using the data above but changing the duty factor to 0.0001, we get:

$$S_{max} \text{ (mWcm}^{-2}) = (0.0001 \times 20\,000)/0.95$$

$S_{max} = 2.1$ mWcm^{-2}. This is only 10% of full scale. Care would be needed to avoid sudden damage to the probe if the reading increased above that figure.

Similar calculations for probe B (f.s.d. = 100 mWcm^{-2}) using duty factors of 0.001 and 0.0001 and a calibration correction factor as for probe A are respectively:

For DF = 0.001: $S_{max} = 315$ mWcm^{-2} i.e. instrument fully useable to 100 mWcm^{-2}.

For DF = 0.0001: $PD_{max} = 31.5$ mWcm^{-2} i.e. restrict to approximately three tenths of f.s.d.

If we take the ratio of the peak power density rating to the greatest range f.s.d. and term this the peak power density safety factor (SF) for an instrument, we get, for probe A:

$$SF = \frac{\text{peak power density rating}}{20\,\text{mWcm}^{-2}} \text{ (both in same units)}$$

SF = 20 000/20 = 1000 (30 dB if decibels are preferred)

and for probe B,

SF = 300 000/100 = 3000 (34.8 dB)

Figure 8.1 shows a plot of four values of SF against permitted highest range meter reading and pulse duty factor, expressed as a percentage of full scale.

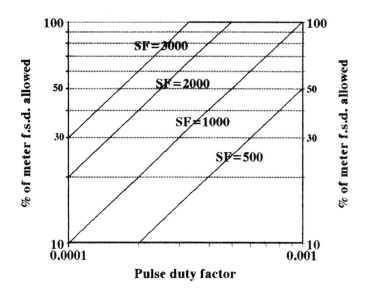

Figure 8.1 *Avoiding meter damage–plot of four values of safety factor (SF) showing safe meter readings against pulse duty factor*

It can be seen that where the duty factor reciprocal exceeds SF, some reduction in the maximum permitted reading will be required to avoid sensor burn-out. Figure 8.1 cannot reflect the calibration correction factor of an instrument so that the percentage of full scale extracted from it should be divided by the calibration correction factor.

Because these parameters may be expressed in different ways by manufacturers, the handbook should be consulted in any specific case. For example the peak overload limit is sometimes given in terms of pulse energy density i.e. $W\mu scm^{-2}$. In this case it is necessary to divide the limit as expressed by the pulse width in microseconds to determine the relevant peak pulse power density.

Measurement problems

A useful paper from the Narda Company[46] highlights a number of problems which can occur when using RF radiation meters.

1 Magnetic field measurements in high impedance fields

When using a magnetic field probe in a field where most of the energy is in the electric field, a negative reading may occur. This is attributable to the deposition of energy in the resistive transmission lines in the probe. If there is a reasonable amount of magnetic field energy producing an upscale reading this effect would be negligible against the upscale reading.

If most of the energy was in the electric field, a small effect would not be of significance unless there was a definite downscale indication. This latter effect would suggest that the wrong field is being measured from the point of view of identifying a potentially hazardous field.

2 Transmission line antennas

At low frequencies, below about 1 MHz, the change of impedance of the sensor antennas may result in the transmission line delivering an induced signal to the sensor and thus produce an indication which is greater than it should be.

3 Diode light sensitivity

Instruments using hybrid Shottky diodes need suitable provision to prevent them responding to light and to infrared energy.

4 Potential field effects

At low frequencies, usually below about 1 MHz, when measurements are made close to the radiator, a false reading may result from capacitive coupling into the scalar field. One method of reducing this effect is to put the meter and probe in a zero potential field near the ground. Another method is to bring the probe and meter close together by clipping together, if this is possible and isolating from the ground by using gloves, thus elevating the meter unit to the same potential as the probe.

A useful check on whether the readings are in error or not is to cover only the sensor area of the probe with cooking foil and measure again. If the readings are little affected by the foil, they are likely to be erroneous. It is important that the foil does not contact earth, the meter or the rest of the probe.

5 Static field effects

This has already been mentioned in connection with leakage measurement as a source of erroneous indications, coupled with the danger of getting in the habit of disregarding indications thought to be due to static charges, which might really be due to the RF field. Some plastic materials including plastic file and report covers can generate large amounts of static, adding to the general nuisance.

Planning surveys

Nature and purpose of surveys

The main reasons for undertaking surveys include:

- New designs.
- New installations.
- Changes to installations (power, frequency, beam characteristics, etc.).
- Structural changes on site (buildings, portable cabins, towers, etc.).
- Safety audits and routine safety reporting.
- Changes in legislation or safety limits.
- Alleged over-exposure of people.
- Anxieties expressed by the people employed or by visitors.

Some of the above, such as new designs, modifications and updates, are likely to involve the designer in surveys before the user becomes involved. For the others such as structural changes on site and changes in installations of a local kind, the user will have the primary responsibility to initiate further surveys.

It is impossible to generalise on survey planning since the amount of work varies according to the nature of the task and may range from almost none for a survey of a small item which is well known to the surveyor, to several days of work for the full survey of a complex site which has many transmitters. There may also be difficulties in getting transmitters out of service, especially in the military field and in broadcasting. It is not unusual for surveys to have to be done progressively in a number of time slots in between periods of operational use. This applies particularly in the HF broadcasting field.

Similarly a survey of a location which uses a variety of RF process machines or RF medical equipment may involve appreciable planning work according to the technical and organisational factors applying. For example, it may involve detailed planning to arrange the availability of the equipment to be tested and to ensure that any interfering equipment may be switched off. In the case of process machines it may be necessary to arrange the availability of suitable work pieces so as to test machines in a representative state. It can also take time to find the necessary technical data on machines, as handbooks easily go astray.

Tables 8.2 and 8.3 provide simple checklists for leakage and exposure surveys. In practice, surveys often take in both aspects. This is particularly the case when new equipment is being commissioned. The checklist tables should be treated as a starting point since, for greatest effectiveness,

Table 8.2   *Leakage surveys checklist – transmitters and RF plant*

---

1 *Requirements of survery*
   Purpose of survey and use to be made of the results; Statement of what is to be measured; Whether X-ray and/or RF to be covered.
2 *Nature of sources*
   Type and power rating;
   Frequencies involved;
   Type of modulation/pulsed transmission;
   Authorised operating conditions; prohibited conditions;
   Operational or other limitations on the availability of the equipment.
3 *Ancillary equipment involved in the survey*
   Dummy loads, feeders, aerial exchanges, etc.
4 *Availability for testing*
   Times available;
   Availability of ancillary equipment.
5 *Previous reports*
   Reference numbers of any previous reports;
   Details of any changes in layout, equipment replacements etc., since the last report.
6 *Safety standards*
   Standards to be used to determine report recommendations.
7 *Safety management*
   Relevant details of existing radiation safety management practices.
8 *RF plant only – machines and process equipment*
   Arrangements for workpieces and operator to be available and for interfering machines to be closed down.
9 *Personnel safety*
   Any special hazards to survey personnel other than those related to the RF and X-ray measurements being undertaken e.g. hazardous substances in use.

Table 8.3   *Checklist for exposure surveys – transmitters and other sources*

1 *Requirements of the survey*
 Purpose of the survey and use to be made of the results;
 Extent and details of the measurements required.
2 *The site*
 Location; map data and topography; drawings of site layout;
 Relationship to highways and public footpaths;
 Any sharing of the site by people not employed there e.g. sub-letting to farmers, etc.
 Explosives and flammable substance stores; Helicopter operations.
3 *Nature of the radiating equipment*
 Types of equipment and antennas;
 Dimensional and azimuth/elevation angle data for antenna systems;
 Antennas to be used, if there is a choice;
 Heights of towers and mounts;
 Frequencies, powers, types of modulation;
 Authorised operating conditions including operating restrictions and 'permit to work'
 arrangements;
 Limitations likely at the time of survey e.g. 'radio quiet periods', unserviceability; service
 demands, etc.
 Human activity related to the equipments e.g. rigger access, tower climbing, etc.
4 *Off-site environment*
 Nearby gas terminals, petroleum/oil installations; residential property close to site, mobile
 radio transmission, etc.
5 *Previous reports*
 Reference numbers of any previous reports;
 Details of any changes in layout, equipment replacements, etc., since the last report.
6 *Safety standards*
 Standards to be used to determine report recommendations.
7 *Safety management*
 Relevant details of existing radiation safety management practices.
8 *Personnel safety*
 Any special hazards to survey personnel other than those related to the RF measurements
 being undertaken e.g. hazardous substances in use.

checklists should be compiled locally to take in the specific features of the organisation. People are much more likely to follow procedures which they have helped to create.

Unless the survey is being carried out at a location with which the surveyor is familiar, for example in a part of the surveyor's own company or organisation, it can be desirable to obtain the necessary information in a written form as a response to a particular questionnaire, since verbal responses are often not well thought out and can mislead.

Also the person making the verbal response might not be the right person to answer since he may not be aware of all relevant aspects, whereas a written answer is likely to be subjected to further scrutiny and possible correction by management.

This is most likely to apply to exposure surveys where questions about the presence of flammable substances and electro – explosive devices and related aspects can be of considerable importance.

Leakage surveys will usually be easier than exposure surveys from the point of view of acquiring the necessary information, since less information is involved compared with that for antenna systems. Also the risk to the surveyor is usually less, given reasonable care, since leakage from equipment is limited as to range and potential.

Exposure surveys, dealing as they do with intended radiation, can generate many more questions extending to possible irradiation of the public in the public domain, possible hazards to aircraft, flammable substances, medically-implanted devices such as heart pacemakers and other specialised devices.

For complex surveys, it can be very helpful if the location can be visited before the survey and both the supervisor and the equipment operator contacted. If production plant is involved, then understanding the sequence of operations, the RF power duty factors for each machine and any variation with different types of workload may be very important in making any judgements necessary.

However distance, cost or excessive workload may preclude this, in which case as much information as possible must be obtained by the use of a checklist questionnaire.

Equipment and site topographical data

For any site the key information required is that related to the topography of the land, details of the surrounding area and any known special risks such as petroleum installations, gas terminals and the like.

There is also a need to know about antennas and structures, including buildings, their dimensions and spatial relationships. This is likely to include:

Microwave antennas

These may include fixed dish antennas, moving antennas such as surveillance and height finding radars, tracking radars, and satellite dishes. The data required is: the transmitter and antenna parameters: frequency, power, antenna dimensions, gain and beamwidth.

Positional information: height of antenna centre above ground, bearing of fixed antennas, beam elevation angle including range of possible adjustment of elevation positive and negative relative to the horizontal.

For antennas capable of moving, the azimuth and elevation scan angle, sector blanking angle relative to a reference, if applicable; scan duration and rotational rate for continuously rotating antennas. The height and angular information required is illustrated in Figure 8.2 and Figure 8.3.

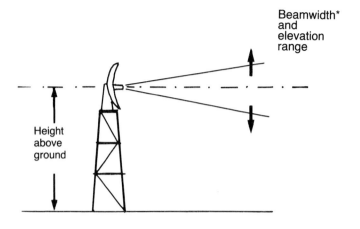

Figure 8.2 *Height finder height data and angle data*

*Azimuth and elevation beamwidths

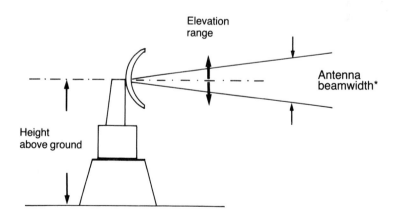

Figure 8.3 *Surveillance radar height and angle data*

Other antennas: type gain and height above ground

Heights on ordnance survey maps or or maps derived from them are relative to the survey datum. It is often easier to use a local datum to improve scaling resolution, for example, by making the lowest ground height on or near the site, zero on the diagram. (See Figure 8.4). In the absence of map data it may be necessary to use an optical instrument to determine the relative heights.

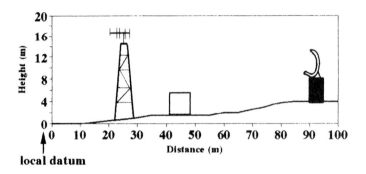

Figure 8.4 *Depicting relative heights in elevation diagrams*

Personnel on site

Exposure surveys on large sites or on sites having a lot of transmitters can involve a great deal of trigonometry to ascertain which antenna might pose a hazard at places where people work. Since equipment mounted on towers has to be maintained, the maintainer will, on occasions, require to climb those towers and this must be allowed for as well.

In order to take into account the exposure of people on site it is necessary to establish the pattern of personnel movement on a site. The sort of interrogatives much loved by those who run management training courses do turn out to be useful in defining the potential human exposures:

Who has to work on the site?
Where do they have to work?
Why do they work there – maintenance, operation, administration, security, etc.?
When do they work there – day work, shift work, on demand?

In certain situations the last question can be very important. For example on a high power HF broadcasting station which has shift manning, it may be necessary to allow for the fact that different frequencies and antenna systems are used at night and these might need to be surveyed as well. This depends largely on whether there is any liability to outdoor exposure near antennas at night.

Generally speaking, activity in antenna fields in the dark with limited lighting can be particularly hazardous in terms of physical injuries such as tripping, contact with sources of RF shock and burns and the hazards of night climbing. It is therefore important to ensure that no unnecessary RF radiation exposures can take place in such circumstances.

Planning and documentation

In the detailed planning work, it is very useful to prepare a rough sequence chart for the activities so as not to lose track of what is needed. The aim is to plan work so that priority services which have been shut down for the survey receive highest priority in the work to minimise their downtime. This may even mean having an inefficient plan in the sense that by doing all the actions on the priority equipment at once, subsequent work on other equipment may take longer.

Nevertheless it is often essential to minimise the dislocation of operational installations and the survey sponsor will have to balance the operational commitment against any consequent increase in the time taken by the survey staff to do the job.

Radio communication is usually essential on large sites, both to secure operational changes and also to be told if a transmitter has tripped off.

Ensuring the availability of communications equipment is thus important. With small transistorised mobile transmitter-receivers it is necessary to avoid accidentally exposing them to very high field levels as the receiver is unlikely to benefit from the experience.

When planning surveys which may have some legal significance or which may be used for planning applications and the like, it is advisable to take particular care to ensure that the sources of radiation are set up at the powers and antenna settings claimed by the equipment operator, through direct witnessing of the activities connected with the transmitters by another surveyor. Again, radio communication is useful for this activity.

The purpose is to ensure that any parties involved in planning complaint investigations and similar activities with legal or social implications can feel that the surveyor really had a full knowledge of the situation and would have spotted any failing on the part of the operator of the source.

After all, it is not surprising that a complainant who feels strongly about a situation may, when shown measurements which have been made and which confirm that a situation is not hazardous, challenge some obvious targets – the validity of the RF source data, the calibration status of the instruments and the competence of the surveyor.

Where, in such investigations, the main finding is that no readings can be obtained with RF radiation instruments, it will sometimes be necessary to resort to more sensitive equipment such as a field strength measuring receiver in order to provide a more definitive answer.

This is mainly a psychological problem where non-technical people do not understand range limitations on instruments and disbelieve the survey results. This is understandable if, for example, interference such as a radar pulse train can be heard on a domestic receiver at a place where the RF radiation meter cannot measure anything! Even some radio engineers show the same disbelief about safety if their car radio is affected by such signals when entering a site!

It is important to collect and marshal survey data in an effective way. Whilst simple surveys of perhaps one or two sources may make limited demands in this respect, surveys of complex sites may provide so much data that it is not easy to see what to do with it. Some sort of methodology is required. The following paragraphs attempt to illustrate possible methods of using the data to determine the measurement sequences required.

It is useful to take a surveillance radar and a height finding radar of the simplest types as examples which encompass most of the likely characteristics of moving beam antennas. For those unfamiliar with radar antennas, a brief reminder may be useful. Chapter 5 deals with the topic in more detail.

Surveillance radar is somewhat analogous to a rotating searchlight in that the beam rotates at a constant speed, typically 6 rpm. The beam can be preset in elevation above or below horizontal. Because the beam elevation, positive or negative, is preset, it is necessary to know the range of setting permitted. For example, it is important to establish whether the elevation may be set to a negative value. If so, the implications for human exposure may be significantly affected.

Because the beam rotates, it may result in RF radiation exposure over the full active scan range, compared with a fixed microwave antenna which has a limited angular range of potential exposure. Constantly rotating antennas can be subjected to six minute averaging.

A height finder antenna has the same possibility of azimuth rotation and also scans in elevation. The elevation may be positive or negative relative to horizontal. Importantly, it differs from the surveillance radar in not rotating at a constant rate. In fact it moves in azimuth and elevation quite unpredictably as far as the observer is concerned.

Surveillance radars, or indeed any antenna moving in azimuth can be blanked over a sector of its scan so that radiation stops in that sector and no hazard is possible. Figure 8.5 illustrates this. Where sector blanking is used, then for surveying, the blanked sector data is required in terms of the limiting angles relative to North or to some other reference which is documented on the site map.

North is usually grid north if a national grid mapping system exists, so that sector blanking data in suitable form can easily be related to such maps.

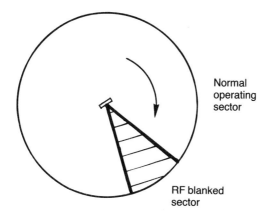

Figure 8.5 *Sector blanking illustrated in plan view*

Note that the centre line from the antenna, the antenna axis, corresponds to the centre line of the antenna charts in Chapter 5 (or any other calculation method) and can later be annotated with the calculated power densities, if desired.

Figures 8.2 and 8.3 earlier, indicate pictorially the height and angular data required for surveys for fixed and moving beam systems. Additionally for the height finder, it is necessary to know about the parking characteristics – some 'park' pointing downwards producing a particularly hazardous region at ground level whilst others park downwards but switch off the RF when not active, sometimes switching off after a set delay.

Simple site example

Figure 8.6 is a simple site diagram showing a site enclosed by a fence, a surveillance radar (S), a height finder radar (H), two buildings, (buildings A and B) and a fuel store (F). Their use in this example is solely to illustrate the amount of investigation needed as a result of the antenna movement.

This diagram is the simplest case, a flat site. In a practical site, either spot height data or contours need to be shown and may significantly affect the safety issues. The two most basic issues within a site will be:

Man-clearance

This is a well established term for ensuring that people can safely walk around the permitted areas of the site, especially near to equipments. It is not intended to be confined to males and can be read as 'personnel clearance'. The clearance height is derived from male height statistics and is usually

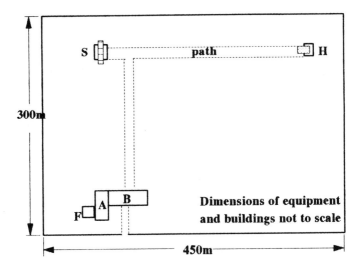

Figure 8.6　*Simple site layout illustration*

taken as being a height of 2.5 m (8 ft 5 in) which is probably the highest the average man can reach whilst correctly operating a survey meter, without having to stand on something.

It means measuring all the possible places where people are allowed to walk and establishing that the power density or other field quantity, as appropriate, does not exceed the permitted limit.

Normally, footpaths or roadways exist to the various equipments, since vehicle access is generally involved. This is obviously a primary area for man clearance, and the paths are shown in the diagram by dotted lines.

Irradiation of other objects and buildings on site

The objects in the diagram have been limited to the two radars. Each radar will be an object to the other, in the sense that each may irradiate the other if the relative heights of mounts result in this situation. Each radar may, in principle, irradiate buildings A, B and F. It is therefore necessary to investigate:

1 Possible damage to one radar by another.
2 Possible irradiation of personnel working on one radar if the other is still operating e.g. during maintenance periods when one is out of operational service.
3 Any irradiation of the buildings and the people associated with them. Additionally, any effects on things or substances in the buildings may be

of importance. Ground access to the buildings will be covered by the man-clearance measurements.

4 Irradiation of people outside the fence.

Item (1) above is only likely to need looking at if the equipment is new or something has been changed, since if the equipments are operational there is obviously no damage occurring! (4) above depends on what is outside the fence and is not pursued here in this example in order to keep it simple.

The task now separates out into three sequences covering man-clearance including access routes, human safety in the buildings and the safety of substances and things in the buildings. Figure 8.7 shows a line for each

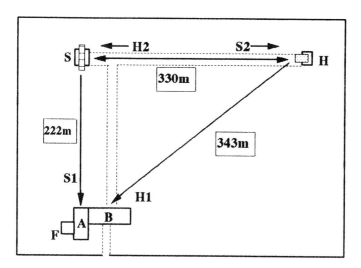

Figure 8.7   *Figure 8.6 with added lines representing test alignments*

required test so that the line S1 from S to the buildings means that tests at A, B and F are needed in respect of radar S, etc. Assuming that tests are being done with the radars stationary and successively aligned along S1, S2, H1 and H2, the measurement requirements specified above, other than those on the footpaths to the equipments, should be covered.

It will be necessary in addition to align the radars to check points on these paths, except where other measurements show that there could not be levels above the permitted limit on the footpaths.

It is not meaningful to try and illustrate the latter in Figure 8.7. In practice, in order to limit the total number of alignments of the radar, the lines shown such as S2 and H2 also permit measurement along the pathway between the two, and the alignment of H1 could be re-orientated in steps to the H2 alignment making measurements on the pathway between S and the buildings at each step.

Table 8.4 *Illustration of a site test schedule*

Test no.	Source subject	Alignment	Measurements
1	S	S2	1 Measure along path S to H for man clearance.
			2 Measure at H to represent maintenance condition.
2	S	Step clock-wise from S2 to S1	3 Check along path S to A and B for man clearance at each step
3	S	S1	4 Measure at the buildings A and B
			5 Measure at F against safety limit for flammable substances
			6 Measure around the buildings for man clearance.
4	H	H1	7 Measure at the buildings A and B
			8 Measure at F against safety limit for flammable substance.
			9 Measure around the buildings for man clearance.
5	H	Step clock-wise from H1 to H2	10 Check along footpath A, B to S for man clearance at each step.
6	H	H2	11 Measure along path H to S for man clearance.
			12 Measure at S to represent maintenance condition.

The same applies to the S2 alignment being stepped round to the S1 alignment. This leads to a schedule of tests as shown in Table 8.4.

It can be seen that in this simple case, at least 12 sets of test measurements are needed, in principle. If there were other objects on site which involved people, the number of tests would increase considerably.

However, in any particular case, some tests may prove not to be required because a beam is aligned well above anything involving people or because antenna calculations show levels to be extremely low. Again, if both radar equipments are always taken out of service for maintenance together, tests 1(2) and 6(12) may not be required. What is most important is to plan for all the possibilities and then eliminate what is unnecessary on logical grounds or as the result of measurements and not on calculations alone. The remaining test requirements can then be planned in an efficient sequence.

Note that Table 8.4 assumes that the flammable substances store is empty! If not, the source should not be pointed at F but aligned in a representative situation e.g. towards the bottom right corner of the site and representative measurements made at a distance from the source which is the same as the distance between the source and F.

Even if the result shows a good safety margin below the hazard level, the beam should not be pointed at F but an allowance of say 6 dB should be made for possible local enhancements and the resulting figure regarded as the value to be assessed against the relevant standard. It is obviously preferable to remove flammable substances to a safe place elsewhere during testing to permit direct assessment.

Although this is a very simple example, it can be seen that by careful planning, the efficiency of the survey can be improved and and the time duration of the reduced. Also, the nature of the planning can be varied according to the operational situation. In the example given, it would be necessary to get a small overlap when both radars were out of service so that either alignment S1 or H1 could be done without harm due to the surveyor having to stand at or on each radar in turn, to simulate maintenance.

If the two antennas were fixed rather than rotating beam systems e.g. communications systems, the testing would simplify to measurements around each, though the interest in possible elevation and azimuth angle setting changes would remain.

A flat site is fairly rare, and it will often be necessary to draw detailed site sections to look at the height of an antenna relative to another or relative to buildings. Figure 8.8 shows an elevation view of the site to get some feel for

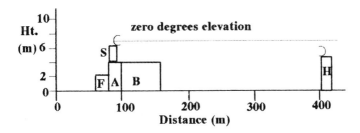

Figure 8.8 *Site sectional diagram in elevation*

the height relativity. As it is a flat site it is effectively a section of lines S2 and H2, if the buildings are ignored where irrelevant for a particular line.

It would be useful to plot the individual lines S1 and H1 in order to see the relationship between the two points being examined including the effective distances between the radars and the buildings.

It can be seen from the distances on the diagram, S is only 222 m from the buildings whereas H is 343 m from them. In drawing Figure 8.8, it was obviously necessary to assume some values for the antenna heights above ground and the building heights. The surveillance radar beam elevation angle is drawn as zero degrees.

It can be seen that if, for example, radar S is mounted reasonably high up, as it is in the diagram, it may be the case that it will not give any problem on site unless it is set to a negative elevation. If it is mounted on a very low mount, then it is very likely to give problems. Whether there is a problem or not will also depend on distance from the antenna to the place affected, the transmitter mean power and the antenna gain.

Where individual ground heights at specific equipments vary because a site is far from flat, then these height differences may affect the interaction between antennas and buildings adversely or advantageously according to the siting of equipments.

This example is also oversimplified in terms of the equipment on site, since on a real site many other equipments may be present e.g. communications systems, other radar equipment and so on. Indeed there is an increasing tendency for towers to collect all sorts of antennas, often for outside users, so that many radiations may have to be explored.

Also other objects may be present such as spare antenna parts, mast sections stored on the ground, scrap metal structures and vehicles. If there are any transmitters with frequencies up to about 100 MHz, it may be necessary to check for possible shock and burns from these items. If there is anything similar in the adjacent public access areas which could cause burns, it may be necessary to do a discrete simulation on site if possible before risking causing unnecessary alarm to non-technical people by doing measurements outside! Farm tractors, caravans and other vehicles are often found close to site boundaries and provide significant conductive masses.

The result of the extra complication of real sites is that the schedule of tests to be done will be much more extensive and require more detailed planning and organisation.

Note that when beams are illustrated in diagrams such as Figure 8.3 they are shown as bounded by two lines (the half power limits) for simplicity. However, in power density terms there is appreciable energy outside the lines, i.e., if the power density on the axis is, say, 300% of the permitted limit, 3 dB down is still 150% of the limit!

Hazard avoidance and remedial action

When planning surveys, it will often be necessary to consider what action might be possible in particular circumstances where it seems likely that some safety limits cannot be met. This problem might be seen when studying layout diagrams and the antenna calculations and if it is found to be correct, some recommendation will be needed to deal with the situation.

When actually carrying out surveys, it is often the case that additional shortcomings are identified which stem from close inspection of what happens on site and these are noted and included in the recommendations. They usually involve such things as the lack of suitable safety procedures or the failure to use existing procedures.

The following list is not exhaustive but can be used to compile local lists of possible actions which can be considered in particular surveys, consulting with the relevant people to eliminate those which are incompatible with the operational or other requirements.

Transmission systems

1 Re-siting equipment/antennas

Mobile equipment may lend itself to re-siting; large fixed systems are unlikely to be re-sited except perhaps as a long term action.

2 Use of barriers

Use of barriers to restrict access, including lockable climbing barriers (see later).

3 Sector blanking of rotating radars

Sector blanking (arranging the radar to radiate over less than its normal azimuth range) can be very effective but since radars have defined requirements for azimuth and elevation coverage, blanking is unlikely to be popular in the operational sector. However, it can be very useful when radars are under test, for example on the manufacturer's test sites, since most testing will not require a full azimuth coverage.

4 Elevating beams to clear working areas

Again, subject to operational requirements, applying minimum elevation restrictions can be very effective in dealing with awkward problems. Equally, increasing the height above ground of the antenna can be of similar value where practicable. Both measures can solve the problem of man-clearance to walk round an antenna system.

5 Re-siting of vulnerable 'targets'

Sometimes when the list of other possible remedial actions have been dismissed it may become necessary to move, say, a flammable liquids store to a safer place. Similarly, with the increasing use of portable cabins for technical staff, the moving of such cabins is a possibility. Design engineers, in particular, do seem to have the habit of locating such office accommodation as close to the place of work as possible without much regard for safety.

6 Confining human access

Confining human access to clearly defined paths and roadways which are chosen as safe routes.

7 Shielding of huts and cabins which cannot be moved

This is a rather expensive approach although this has been done successfully in a few cases, using cheap wire netting. However, the fact that an occupied hut has to be screened implies a hazard to any personnel who walk out of the hut, unless the exit is remote from the irradiation due to being located at the rear. Some people use a 'keep walking' provision on the basis that if they keep walking away from the source, they will meet the exposure limits under the six minutes averaging provision. Unfortunately, when two or more people meet they are inclined to stand and talk!

8 RF shocks and burns

The best solution if at all possible is to remove stored metal objects and scrap metal from the RF field.

For objects such as masts, antenna mounts, and other items which, for operational, technical or financial reasons are not movable, these can be surrounded with a rope barrier and signs if in the occupational environment. In the public domain, the problem is that the objects are likely to be owned by the public rather than the RF site company or organisation. Sometimes agreements can be made with the owners or occupiers of the land. Purchasing or leasing of the land involved might be another possibility, although possibly not a very attractive one. Failing this resiting of the antenna or power reduction might be necessary.

With the general tightening of RF safety limits for public exposure, those setting up new transmitter sites which provide potential problems of high exposure levels or a shock and burns risk might well look to the possible purchase or leasing of extra land to avoid future problems.

9 Transmitter power reduction

Power reduction is very unpopular both from the technical point of view and from the human one – why pay extra for a higher powered transmitter and then reduce power? Nevertheless, there are circumstances where this must be done.

For example, if the public, in the public domain, are being irradiated at levels higher than those permitted, then there may be no choice. A similar situation can arise if the transmitter is near a gas plant or a plant storing or manufacturing substances with flammable vapours and it can be shown that these might be at risk.

10 Permit-to-work provisions

Often, a particular hazard can be dealt with by a 'permit to work' procedure.

Unfortunately, procedures controlled by human beings fail regularly unless they include physical controls and are audited regularly. Nevertheless, the system has to be used in circumstances where it seems most appropriate. One of the common failures with 'permit to work' systems stems from the limited communication of the restriction within an organisation so that sub-contractors are contracted to undertake maintenance, grass cutting, building work and similar tasks without any real knowledge of the hazards of RF radiation. A few examples from past experience might illustrate the problem:

(a) The side of an aircraft hangar being fitted with new sheeting by workmen on scaffolding and generally progressing to the far corner of the hangar which was being irradiated by three high power radars on test.

(b) A telephone call from a customer indicating that one of his maintenance staff had just been hoisted up the side of a working tropospheric scatter antenna (10 kW).

(c) A technician sitting in an aircraft radome, unaware that the radar was working.

(d) A technician standing between a dish antenna and the feed, dismantling the antenna whilst it was operating. His back felt warm! He did not speak English, needed two authorisations from the two radio rooms listed on his otherwise well-organised permit to work, but only got one signature – demonstrably the wrong one.

(e) Technical people who also believe that because they understand RF radiation, RADHAZ warnings are not meant for them. They are usually wrong on both propositions!

(f) A subcontractor cutting grass in front of a working microwave dish on a low mount, removed the access prohibition notice in order to cut the grass right up to the dish and then replaced the notice!

10 Associated safety practices

Particularly useful practices include:

(a) 'Key to operate' controls using key switches. The use of good quality switches with keys which cannot be cut in the local shop and are not duplicated on site.

(b) Use of start-up alarm systems to warn of impending equipment starting to move or operate. With the amount of energy available in large rotating or otherwise moving radars the potential for injury is liable to be much greater than that of the radiation, as far as a person walking past is concerned. This is not to underrate the potential hazard of RF but rather to emphasise what is often overlooked.

(c) The use of flashing lights on radiating antennas. The use of flashing lights operating only when RF is on is particularly useful, especially on test sites where the pattern of work may change from hour to hour. With such an arrangement a radar rotating to test the bearings without the RF on would not flash until the RF was switched on.

(d) A novel approach to safety on towers and masts which has been used in Australia was reported by Hatfield of Telecom Australia. Interlocked 'traffic light' style stop and go signals were fitted to the tower and controlled remotely. Only a 'go' light permits access, a stop light or a light failure indicating a prohibition.

RF machines and plant

1 Shielding for operators, particularly for the lower limbs in 'seated at machine' situations. Situations occur where operators have found the nails in their boots getting hot! Steel or aluminium can be quite effective for this purpose.

   Where necessary, the use of transparent shielding materials of a proprietary kind can be considered where it is required to see the work piece. Some experimental work may be required but there are a number of types of conductive transparent material available which may reduce high electric fields.

2 Control of screening panel removal and refitting-often left off after maintenance.

3 Operator safety training using SIMPLE explanatory methods.

4 Prohibition of the use of flammable vapours near RF machines e.g. petroleum, etc. (See reference 31 or a similar standard.)

Signs and barriers

Practical experience shows that whilst some of the foregoing provisions are usually implemented, the method of implementation is sometimes inadequate. The following observations result from practical problems experienced.

Adequate RF radiation signs (accompanied by additional statements where necessary) should be displayed at the access to all areas where the power density (or a field quantity) exceeds the relevant standard. On sites to which the public have access, physical barriers may also be necessary.

As far as the occupational situation is concerned, the ready availability of passive infrared detectors (PIRs) to detect the presence of people offers the possibility of sounding an alarm to warn personnel who have strayed into a field without realising it. This might help to prevent the occasional cases where people get exposed unknowingly, often due to being deep in thought about an equipment problem. An example is the prevention of maintenance people getting into aircraft radomes when the radar is operating.

Where equipment is not operational i.e. during research and development and routine testing, it is possible to use a PIR device to switch off the radiation. They can also be used in conjunction with some of the fixed radiation monitors mentioned in Chapter 6.

No account can be taken of notices or signs where children are involved and measures taken should prevent access. In particular, the climbing of mast ladders should be prevented by lockable physical barriers. The latter is often equally applicable to adults since locked barriers are more reliable than prohibition signs alone, both for technical people and the public (who may not be familiar with signs).

In most countries there is some sort of control of public rights of way and no restrictions or prohibitions can be applied to public rights of way by those operating RF transmitters. Warning notices near such paths and applying to the adjacent land should be clearly worded! Signs which are thought to imply hazards or frighten path users will cause a great deal of unnecessary trouble.

It is desirable that a suitable warning notice should be sited at the entrance to a site or establishment, asking visitors who may have a heart pacemaker to report the fact on entry. Because of the difficulty likely in establishing the limits of safety with these devices, personnel using pacemakers should not be subjected to significant RF fields.

Creating records and reports

General

In the author's view, it is essential to consider the nature of the survey and report well before doing the survey. Surveys have four basic steps:

Step 1 Establish what is to be measured, details of the sources available and the environment in which they are used.
Step 2 Carry out survey.
Step 3 Establish findings; do any rechecks necessary.
Step 4 Produce report.

Records involve the actual measurements made, which may not only include RF and X-ray measurements but also the associated linear distances, heights, and other related details. When planning surveys, thought needs to be given to methods of documenting the results. There are many ways of recording and documenting test results and in very simple survey tests, few problems are likely to occur. However, in large surveys, the problems of recording, converting and presenting results can prove to be a real problem.

If we analyse the requirements we can identify two areas at least where some sort of organisation is needed. These are:

1 Recording measurements and, where necessary, converting the values to those needed for manipulation.
2 Presenting the results to those who initiated the requirement for the survey.

Recording results

Recording results sounds so easy that it may be difficult to see what problems can occur. However, in any survey other than the very simple types, there may be a considerable amount of data to be produced involving:

1 Recording the actual instrument readings and the relevant locations.
2 Converting these values from meter readings to the quantity actually required.

For example, meter readings in $mWcm^{-2}$ on a meter used for electric field measurement will involve the conversion to Vm^{-1}. People used to working in Wm^{-2} are likely to do the intermediate conversion $mWcm^{-2}$ to Wm^{-2} and then convert to Vm^{-1}.

A simple form is the usual arrangement for such recording. The nature of the form can be determined from local considerations but it should ensure that the actual indication of the instrument is recorded before any conversion, as this permits recovery from conversion errors which can easily occur under pressure.

It is very desirable to have a standard practice for units so that people working together use the same arrangement. Preferably this should be SI units, but the most important thing is to have a clear policy. If two people work together, one used to $mWcm^{-2}$ and the other used to mentally converting to Wm^{-2}, experience shows that there is likely to be some confusion over the recorded results.

Surveyors also need to memorise safety levels necessary for them to avoid personal hazards and as most of the main standards now use Wm^{-2} this should be easier (the exception is IEEE 1991). Similarly, since ANSI was the only standards body giving the magnetic and electric field quantities in terms of their squares $(H)^2$ and $(E)^2$ and has now ceased to do so, safety limits in Vm^{-1} and Am^{-1} should not give rise to any confusion.

The likelihood of mistakes in reading the indicated values tends to increase when surveying in adverse weather conditions due to fatigue and reduced attention to the work. It often results from misreading scales despite the colour coding and other good quality practices used by instrument manufacturers to prevent misreading. Most of us are more prone to errors than we care to admit!

Pictorial recording of data

Identifying locations can be a problem. It was noted earlier in respect of leakage surveys that the use of marking tape on the equipment to identify leakages can be extremely useful. Tape markers can be given identity numbers or letters. The report will need leakage measurements from an equipment to be related to a location for example, by annotating a sketch of the equipment or of the working area. Preparation of such sketches beforehand can be a great help as they can then be copied and the blanks marked up during the survey. Generally a few good illustrations are better than pages of words.

Similarly, exposure measurements need to be documented in a way which can clearly be related to the real world in the form of the radiation sources, adjacent buildings, objects, paths and roads. In the case of RF process machines the nearest things may be work benches, other machines and processes , and the people associated with them.

Such machines will usually involve leakage exposures relatively close to the machine. However some modern RF machines and the associated work flow equipment are quite large and cover an appreciable floor area.

The essential thing is to present location information such as diagrams in such a way that those who have to undertake some corrective action know what to do and where to do it. In the case of diagrams relating to exposures, the recipient of the report may expect enough information to make any necessary changes in the location of people or equipment or to see that such changes proposed in the report make sense.

A method of presenting numerical values of quantities and their locations for complex transmitter cabinets which has been used for many years is illustrated in Figures 8.9 and 8.10. Figure 8.9 shows a line drawing of the

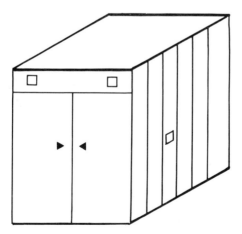

Figure 8.9 *High power broadcast transmitter line drawing*

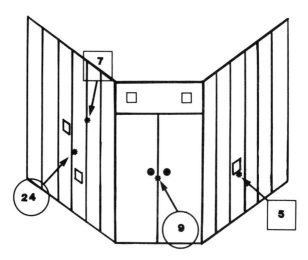

Figure 8.10 *Exploded diagram of Figure 8.9 with measured values annotated*

cabinet layout for the transmitter of a high power HF broadcast installation. The modulator has a similar diagram and is not shown here. The diagram only identifies key items related to possible leakage but can only show one side of a double-sided cabinet bay. Figure 8.10 shows a simple modification to turn the diagram into an 'exploded' picture showing both sides of the cabinet. By an arbitrary convention, RF leakage data is shown in balloons and and X-ray leakage data in boxes.

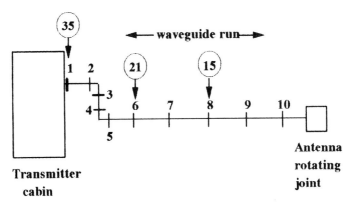

Values in balloons are all W/sq.m.

Figure 8.11 *Method of sketching waveguide runs and the measured leakages*

The leakage points have also been marked on the equipment with an adhesive tape so that it should be clear where remedial work is required.

For waveguide runs the business of drawing can be a time-consuming task unwelcome to the inartistic. Figure 8.11 shows a simple line drawing method which uses a line for each flange and arbitrary numbering of the flanges. It is then possible to go to the right flange and do remedial work.

Controlling changes to equipment and the site

For a site on which relevant points above have been implemented, the main problem likely to be encountered is that resulting from technical or administrative changes. These pose a constant problem and include:

- Equipment changes.
- Power, frequency, feeder and antenna alterations.
- Adding new equipment, masts and antennas.
- Administrative or organisational changes which also need pre-authorisation may include:

  - Erection/demolition of buildings. This may change field distributions or change screening characteristics.
  - Building maintenance, with any climbing and high level work involved.
  - Electrical wiring work in areas of significant RF field.

- Change of use of buildings e.g. creation or re-siting of a 'flammable materials' store.
- Rigging, mast and antenna alteration, maintenance or erection.
- Use of cranes, ladders, etc., in areas of significant RF field.
- Use of low flying aircraft, e.g. helicopters, in RF danger zones. Here the danger may primarily be related to the effect of an RF field on the aircraft control equipment, rather than direct effects on personnel.
- Letting land e.g. for animal grazing, without considering in sufficient detail the safety implications for the people concerned in looking after the animals.

All changes to equipment should be subject to effective control. Activities involving sub-contracting of works is a common cause of safety problems, since those arranging sub-contracts are frequently unaware of the technical aspects of radiation safety. Here again the requirement for a safety signature before sub-contracts are let, will help to reduce problems. It is worth remembering the old adage 'anything that *can* be done probably *will* be done, sooner or later'.

9
Conducting radiation surveys

Part 1 Leakage surveys

Personal safety in surveying

It is obvious that those who undertake surveys have to be prepared for some risks, especially when assessing new designs of high power equipment. Such risks can be made negligible by care in the initial approach to equipment. Whilst some safety points are mentioned again in particular paragraphs because of their immediate relevance, the following warnings are generally applicable.

X-rays

Where possible wear an approved type of X-ray dosemeter badge.

When both RF and X-ray surveys are requested, do the X-ray checks first. Also, even if X-ray checks have not been requested but the surveyor has some concern about the presence of X-rays, then those checks should be done first. Have an RF measuring instrument available at the same time to keep an eye on RF radiation!

The initial approach to any equipment which might conceivably produce X-rays should be a cautious one where measurements should be made at as large a distance away as possible. A sensitive sniffer instrument will assist in this.

Where unexpectedly high readings are encountered, either retreat to a safe distance from the exposure to allow time to think out your next move or switch off the source. Such a source should not be left operating unattended.

RF radiation

The initial approach to any equipment which produces RF radiation should again be a cautious one where measurements should be made at as large a distance away as possible. Whilst it is obviously possible to distinguish the risks between a high power transmitter and a small and trivial low power source, it is usually desirable, unless the surveyor is very experienced, to exercise caution on a general basis, thus avoiding any consequences of erroneous judgements. Particular care should be taken with prototypes of new designs. Note that as well as possible hazards to the surveyor, the RF radiation instrument may also be at risk.

Consider the possibility of RF shocks and burns at the lower frequencies.

Avoid allowing technical interest to reduce your attention to other hazards present such as mechanical devices, electrical hazards, other forms of radiation, hot surfaces, etc.

Specific procedures may apply in certain antenna fields such as those at MF and HF broadcasting stations where the shock and burn hazard may be particularly acute or where climbing is subject to special controls. Surveyors should not work alone in high power antenna fields.

Common electrical safety considerations

When carrying out X-ray or RF tests with panels off and high voltages accessible, the surveyor must be familiar with:

1 The fact that the instantaneous PEAK potential on points can significantly exceed the nominal high voltage supply.
2 The 'jump distances' for high voltages in air, i.e., the air breakdown distances whereby, if a probe or other object is brought too close to a high voltage terminal, the voltage can break down the gap and take the object to the terminal potential.
3 The possibility of an excessive amount of RF exposure being incurred whilst doing X-ray tests with panels off and the chance of incurring an RF arc from a point of appreciable RF potential.
4 The procedure for dealing with stored charge in capacitors when power has been switched off.
5 Open panel work, probably mostly confined to the investigation of X-ray radiation in new designs in the manufacturer's development laboratories and test departments, needs two people present throughout the tests and also requires, where the nature of the equipment necessitates, the use of wooden or plastic guard fences to avoid the danger of stumbling into the equipment.

Many people remove rings and metal watches from the hands when doing open panel work. (Note that plastic-cased electronic watches may not survive exposure to the RF!)

Leakage measurement surveys

The starting point for all survey measurements whether for leakage or for exposure is the existence of a safety standard. For the purpose of illustrating survey methods, it does not matter whether this is an established national standard, a company or organisation standard or a contractual standard produced by a customer. The standard provides the criteria for all decisions during the planning of a survey and for all recommendations made as a result of the survey. Sometimes the manufacturer's own safety standards may be more severe than any of the other documents mentioned.

Where leakage refers only to the structure of a high power transmitter, it is desirable, wherever possible, to use a dummy load on the transmitter output to avoid unnecessary radiation.

X-ray surveys

X-rays must be considered as dangerous and it is therefore important to consider the safety of those carrying out X-ray surveys. For those not experienced in dealing with X-rays, the warnings given above should be studied. The use of a sniffer instrument to do the initial checks will assist with personal safety as well as facilitating the detection of X-rays.

The same precautions should be observed for each equipment, however familiar the surveyor is with it. Specific protective items such as lead glass windows should not be trusted until it has been established that the correct type of glass is fitted. When satisfied that there is no sign of excessive or unusual radiation, the survey can proceed.

X-ray radiation is covered by legal provisions. Individual national regulations and codes of practice may have specific requirements which are not necessarily covered here. Note that this may also be the case when the equipment is for delivery to another country. Problems may arise in such cases if the radiation certification is not in accord with the recipient's national regulations.

Where X-ray measurements on high power transmitters need to be done at a number of frequencies, it is quite usual to proceed in the sequence:

X-ray measurement on frequency 1
RF measurements on frequency 1
X-ray measurements on frequency 2
RF measurements on frequency 2
and so on to the last test frequency.

This method of working reduces the number of frequency changes needed and the time involved compared with doing all the X-ray work first and then starting again for the RF measurements, an important factor where the time to change frequency is significant. It is also essential when having to work on broadcast transmitters and other types of operational transmitters in time slots between periods of service, as it then means that both types of measurement for one or more frequencies are done in a given period and the results can be studied whilst waiting for the next free time slot.

Reasonable ambient levels of RF radiation in the building are not likely to affect X-ray instruments, but in case of doubt aluminium cooking foil held across the apparent X-ray beam or wrapped over the instrument window will, except at the very lowest energies, be transparent to the X-ray but stop any RF present which may be the real cause of the meter reading.

The equipment operating conditions should be those that give the maximum power that is permitted or scheduled for use, and, where appropriate, with modulation, i.e., worst case conditions.

The initial survey should cover all the surfaces associated with the source of radiation, front, sides, back, and top and not be limited to that part of the equipment where the source is located.

This is particularly important in a large equipment where X-rays may be reflected around the cabinet and emerge at some unexpected point. All normal covers and panels should be fitted.

The initial survey should be carried out with a sensitive 'sniffer' meter and the instrument should be held as close to the surfaces of the equipment as possible to maximise the chances of detecting narrow beams. Figure 9.1 shows the use of a sniffer instrument on a high power mobile radar transmitter.

Where leakage is found, the location of the leak should be physically marked with some suitable adhesive cloth tape or other material.

Where a Geiger-Müller tube instrument is used for initial surveys on pulsed or amplitude modulated sources, the maker's instructions should be followed with regard to pulse or modulation repetition rates, otherwise it may be found, in the limit, that the instrument is responding to the repetition rate rather than to the X-ray doserate.

To put this problem in perspective, it should be said that the author has used such an instrument for several hundred surveys over about fifteen years without ever experiencing the problem.

After the initial survey, measurements should be made with an ionising chamber instrument at those places where the leaks were found and, for recording purposes, more systematically at these places which are the source of the X-rays such as the final output stage compartment, and high voltage modulator or other source, to confirm that there are no measurable leaks.

All measurements should be made at the nationally used distance, usually 100 mm from the chamber centre mark but sometimes 50 mm for some particular products such as video display units and similar cathode ray tube devices.

Where, as should be the case, the 'sniffer' instrument is more sensitive than the instrument to be used for actual measurements, it follows that at

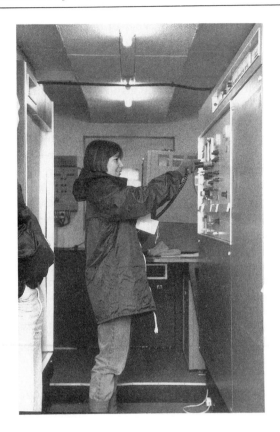

Figure 9.1 *Training course student using an X-ray sniffer on a high power radar transmitter*
  (Courtesy Marconi College)

some of the points which had been identified as having leaks, the ionising chamber instrument may indicate that the beam doserate is too low to resolve. It is desirable to leave such leaks marked but perhaps annotate such marks to indicate that they were unreadable.

The reason for this is that if, subsequently, during the survey some operating condition is found which materially increases the X-ray doserate the leakage points can readily be checked again. In particular, with amplitude-modulated high power broadcasting transmitters further checks may be necessary using high modulation levels (90 to 100%) to identify any rapid increases in X-ray doserate on the lines discussed in Chapter 7. X-ray levels with the carrier only may be unmeasurable in such cases.

Where permitted maintenance includes some work with panels or covers off and the equipment operating, tests should also be done in this condition to safeguard the maintainer's interests. Note that the maintenance situation must not result in an infringement of the ionising radiations regulations. If necessary a maintenance shield should be available, a standard practice with many radar equipment suppliers.

Practical points

Note that the relevance of these points may vary according to whether new equipment is being commissioned or existing equipment is being subjected to a routine survey.

- Suspect door leaks due to contact finger strips and test several times with doors being opened and closed each time. Door catches are also leak sources as they have to have a cut-out for the catch mechanism.
- Glass windows should be lead glass where X-rays are present. Ensure that it is still fitted, in case any local unofficial replacements prove to be ordinary glass! In the absence of any specific indication, it may have to be checked for leakage with power on.
- High voltage RF sources not located on solid ground floors (for example, sited on a first floor level) may need checks in the room below for X-rays (and possibly RF) depending on the nature of the floor-ceiling structure.
- X-ray radiation beams may vary in location when a new electronic tube is fitted, if the electronic tube internal structure has a different orientation relative to the base mounting than the previous one. The problem here is generally one of being too precise and definitive about shielding, designing it perhaps on the basis of a sample of only one or two electronic tubes.

This may also apply if it is possible to fit an electronic tube with a difference in orientation, due to the whole tube being inadvertently rotated. Figure 9.2

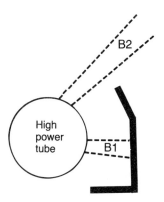

Figure 9.2 *High power transmitter tube showing the possibility of X-ray beams appearing at angles outside the shield*

illustrates both these problems, which are not distinguishable diagrammatic-ally, by a plan view of a new high power tube having been installed with an X-ray screen covering that part of the tube where the designer expected X-ray radiation. In the diagram, B1 is the X-ray beam anticipated by the designer.

A new tube may, either due to internal electrode structures or to the rotation of the tube mount, give rise to an X-ray beam B2 which is not stopped by the shield. The diagram does oversimplify the nature of the X-ray radiation from tubes to illustrate the shielding point.

When it is known that the high power tube produces significant X-ray radiation and the shielding is only partial as just discussed, there should be a leakage check whenever the tube is changed to detect beams not stopped by the shielding. Further problems are possible if more than one type or brand of electronic tube can be installed since there may be differences in the whole electrode structures of the different types of tube, from the X-ray point of view.

Very old electronic tubes may also have less inherent X-ray shielding and can be a particular problem since the equipment designer may not have had any experience of the old electronic tubes or even be aware that any user possesses them. A real case which happened a few years ago, involved a modern radar in a ship which had been fully surveyed for leakage. When a very old replacement tube was fitted, an X-ray beam was found emerging from the radar cabin door.

- Some X-ray sources produce an increase in X-ray doserate with age. Consultation with the supplier may be needed.
- Make sure that the transmitter remains ON when surveying. On large transmitters there is often no indication when working at the sides and rear. It can prove very embarrassing to discover that you have worked for twenty minutes when the transmitter was not radiating and recorded a lot of 'no reading' statements!
- One common problem, mentioned earlier, which can occur on high voltage transmitter surveys is false readings (RF and X-rays) due to the presence of a high static field causing a build-up of charge which gives a meter reading. This will drain away if the instrument is held quite still for a few seconds. Some types of paint finishes on panels may assist the static build-up.

If the instrument has a metal part which can be touched when the instrument is held, it is sometimes a help to touch with the spare hand an earthy point on the equipment under test (e.g. panel screw head) to bleed the static charge away. This technique should NEVER be used when any panels are removed and hazardous voltages exposed since earthing one hand is potentially an aid to electrocution!

Requirements for X-ray leakage measurement in national regulations

Some national regulations identify certain types of equipment such as oscilloscopes, video display units and television receivers, items which may form part of transmitter installations.

These items may be subject to measurement at 50 mm from the surfaces of the item and in order to secure some degree of exemption from some, but not all, of the relevant regulations for 'radiation generators' the radiation doserate limit is specified. In these cases the specific regulations should be followed.

RF leakage testing

There is a great deal of commonality between X-ray and RF leakage testing but they have been presented separately since some people may never be involved in the former. For RF testing, it is assumed that an isotropic radiation meter is used and correctly operated as discussed in Chapter 8.

When starting measurements, with the transmitter or RF machine under test not radiating, check the ambient RF field in the building, if any. Note this for later use as it may limit the lowest reading which can be resolved. If it is too high, it may be necessary to have other sources switched off. Note that X-rays, at the levels likely to be present, are not likely to affect RF radiation measuring instruments.

The initial survey should cover all the surfaces associated with the source of radiation and not be limited to the part of the equipment where the RF source is located. All normal covers and panels should be fitted.

The equipment operating conditions should be those that give the maximum power that is permitted or scheduled for use, and, where appropriate, with modulation on to give a worst case situation. If there is a lot of leakage on an amplitude modulated transmitter it may be worth working with the carrier only and repeating with the modulation on later.

Measurements are normally made at 50 mm from the distance the equipment surfaces unless a standard or specification is invoked which specifies some other distance. Even if a greater distance is specified it can still be advantageous to do the initial survey at 50 mm since it is exploratory and is more likely to identify leaks.

Many RF radiation measuring instruments have a plastic ball on the sensor and this ball has a radius of 50 mm so that the required spacing implies the ball just touching the equipment surface. Other instruments have a plastic cone or a ball and cone combination which provides the same spacing. In practice, the best spacing is just NOT touching the equipment being surveyed as this tends to reduce the static charge problem mentioned earlier.

The quantities measured should be those specified in the relevant standard and will involve the separate measurement of the electric and magnetic fields up to the frequency limit in that standard (e.g. 30 MHz for the current NRPB GS11 1989 recommendations, 300 MHz for the IEEE C95.1–1991 standard). Above the appropriate limit, power density will be measured.

The points where leakages are found should, as in the previous case of X-rays, be marked on the equipment under test. After the initial survey, measurements should be repeated more systematically at those places where leaks are found and also where they are most likely e.g. the final output stage compartment.

Where feeders and antenna exchanges are involved, measurements should similarly be carried out. Figure 9.3 shows training course students measuring

Figure 9.3 *Training students measuring leakage along the waveguide feed of a high power radar transmitter*
(Courtesy Marconi College)

leakage from the waveguide feed from a high power radar transmitter. In the picture they have reached the antenna end of the waveguide.

On multiple transmitter installations it is important to ensure that the correct feeders and antenna exchange parts are tested as it is easy to become confused when working on remotely sited items. Dummy loads may also have to be tested. A dummy load usually consists of some form of shielded resistive element simulating the normal transmitter load and capable of dissipating the full power output.

For factory testing of HF transmitters, the element may be immersed in a demineralised water flow, power being measured from the rise in water temperature. Most dummy loads are generally reliable from the point of view of the effectiveness of the shielding but those dissipating very high powers might be very hazardous if there was a loss of shielding.

Some specialised loads for power testing use slightly different calorimetric principles to measure RF power by comparing the heat generated by the RF with the heat generated by a known AC mains supply using circulating fluids in contact with the dissipating element.

These and other water flow systems usually rely on a connection to the transmitter control system to provide an overriding safe shut down if something untoward occurs such as a loss of fluid flow which might lead to a catastrophic failure. This is in addition to any in-built safety devices in the calorimeter system. The surveyor needs to be aware of this when testing such equipment for RF leakage and ensure that the transmitter connection is made.

Where permitted maintenance includes some work with panels or covers off and the equipment operating, tests should also be done in this condition to safeguard the maintainer's interests, as discussed earlier in connection with X-rays.

Some of the practical points listed above are also applicable to RF leakage. However, in the case of windows, lead glass is not relevant. Where leakage may occur, normal glass will usually be used with a wire mesh of suitable mesh size behind it, or sandwiched between two layers of glass. Any lead glass windows fitted should have wire mesh as well if there is RF leakage.

On very high power transmitters in the MF and HF bands, windows near the final RF output stage without wire mesh fitted can cause catastrophic damage to an instrument sensor as the surveyor scans towards the window. There may be no significant leakage until the glass is reached where the leakage field can become large.

In one particular case on a 500 kW MF installation a magnetic field probe read zero over all of the front panel. At the first approach to the window edge, the sensor was blown up so rapidly that the surveyor had no time to react. Consequently, in such cases the window should be explored first, approaching from a distance with the instrument set to its highest range! Windows are important since people put their eyes close to them in order to observe electronic tubes.

Some problems can occur round doors where contact strip earthing is fitted. On high power HF broadcast equipment arcing can sometimes occur when the strip has not been fitted and set properly or where it has been distorted. This may generate enough heat to oxidise the surface and impair the contact.

On microwave equipment where the contact strip dimensions have not been selected correctly, resonances may occur and cause leakages, instead of stopping them.

National or other test requirements

Some particular products such as microwave ovens have detailed test requirements which have to be followed. For example, a British standard BS 5175:1976[47] has a list of tests to be done including the use of a 'dummy load' of a specified volume of saline solution. There are also tests of the interlock switches, etc. In such cases the requirements of the document concerned should be followed. The USA also has specific tests for these items[48].

Other leakage test methods

Sometimes, instead of the usual leakage tests, purchasers specify requirements for transmitter user RF safety in terms of the exposure of a notional person at a specified distance from the transmitter e.g. 1 metre.

This is usually located at the place where personnel will work - in front of the transmitter and possibly at the rear or even all round. This is illustrated for one face of the transmitter in Figure 9.4.

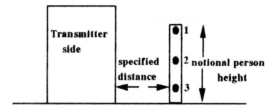

Figure 9.4   *Exposure due to leakage-another method of specification*

A common method of defining such exposure is to require measurements to be made at, say, three heights in the plane of a standing person at the specified distance. These are then added and divided by three to give an average exposure which is subject to a specific limit.

Note that since this measurement has to be repeated at as many places as needed along the length of the transmitter front, rear and sides as appropriate, it will usually be done by initially measuring along the length of the transmitter front, back and sides, at each of the specified heights in turn, in order to establish the amount and general location of any leakage. This will enable a decision as to whether any rectification action is required before taking formal measurements for the certification of the equipment.

An alternative sometimes met is again to define a distance and limit the exposure directly by requiring that the actual power density (or the appropriate field components) does not exceed a specified value at that distance at any height up to some figure such as the 'standard man' height (1.75 m). The only difference here is that there is no averaging, every measurement being required to meet the specification.

The measurement is usually done by holding the probe at the specified distance from the transmitter and measuring in the vertical plane up to the height limit required. In practice, since it is assumed that the person can be at the specified distance anywhere round the transmitter, it is necessary, as before, to survey horizontally along the required length e.g. the front of the transmitter, etc., and at all heights up to the specified limit, perhaps using a suitable number of discrete heights to represent the vertical element of the survey.

RF machines and other sources

It is difficult to generalise about the measurements on RF machines since there is an enormous variety of types and the applications mentioned in Chapter 2 illustrate the wide range of uses. The testing required will be leakage testing but in manually operated machines, operator exposure will also be involved. To avoid duplication, both aspects are covered here.

The frequencies used are those allocated for Industrial, scientific and medical uses (ISM bands) and range from HF to microwave frequencies. These ISM frequency allocations vary according to the part of the world involved. In the UK they are shown on the radio frequency allocation chart[49] as narrow bands without any indication regarding authorisation for use. The internationally listed centre frequencies for ISM use, subject to national authorisations, are 13.56, 27.12, 40.68, 433.97, 915 (not UK), 2450, 5800 and 24.125 MHz. A wider range of frequencies is in use on existing machines, presumably for historic reasons.

In the medical field a considerable number of different types of RF machine are used to apply electromagnetic fields to the human body for therapeutic purposes and for surgery.

In scientific research, many RF machines used have the intended RF output confined to a chamber, sometimes so that it can be pumped down to remove air. In these cases, there is no intended external exposure and only leakage testing is required, as for a radio transmitter.

Processing machines such as those used for dielectric and induction heating in food production, plastics welding, wood glueing and other fields may have leakage associated with the electronic equipment cabinets which generate the RF energy. These need leakage tests both to protect the operator and any other workers nearby. At the work piece there may be a need for exposure measurements as the hands, legs and possibly other parts of the body may be subjected to the fields around the RF applicator in the course of loading and operating the machine.

Since the duty factor for RF machine operations of the production type, that is to say the period of time in an operational cycle when the RF power is on to the total time of the cycle, is relatively small then exposure averaged over the permitted averaging time of any current standard will be appreciably reduced by this factor. A factor ranging between 0.1 and 0.2 may serve as an example but there can be a much wider variation than this in practice.

NRPB report R144[67] gave data from the USA, Finland and Sweden on this topic. The electric field strengths recorded at various parts of the body ranged from about 100 to 900 Vm^{-1} and the magnetic field from about 0.2 to 3.5 Am^{-1}. The figures relate to different machines and different powers but only one table, that relating to Sweden, gives the power information (3 to 6 kW). The date of the report is 1983 and the data may not be true for newly designed machines. However many machines remain in service for a very long time. The report also gives measured field levels for many radio transmitters.

A protocol for the measurement of RF radiation from RF machines was issued by the UK Health and Safety Executive in 1986 as document PM51[50]. The limits used were based on the 1981 recommendations of the American Conference of Government Industrial Hygienists (ACGIH).

The measurement plane is a vertical plane 15 cm away from the machine at the positions of nearest whole-body approach to the applicator or electrodes. The highest and lowest physical measurement heights in this plane are determined by whether machine operation is undertaken seated or standing.

This particular approach is not very useful for many types of machine where the mechanical configuration makes it seem inadequate to use the vertical plane approach alone, as it may not have much relationship to the operator's body and hand positions when using the machine.

It seems more sensible to measure at the places where parts of the operators body come closest to the machine applicator field during use of the machine.

There is more than one opinion on this matter but the author's view is that measurements should also include measurements at the operator's body including hands, etc., in representative operation of the equipment since only then do the full body movements come to light. On machines which are not directly operator fed, for example those which use a conveyor feed, such measurements will not be necessary.

A current approach to the drafting of an EEC standard for the safety of machines by a CEN committee, is to define an effective reference volume for that part of the machine constituting the RF source. A 'measurement surface', defined as being a rectangle, the sides of which are parallel to the reference volume is then postulated and measurement points are located on that measurement surface.

Further measurement points are defined in relation to the working positions of the operator and other measurement points are specified where leakage may occur. When developed this may provide a somewhat more systematic method which takes in some of the points made above. The draft proposes 'type tests' and 'routine tests'.

Of course methods which involve measurements in planes at a specified distance from the source can be useful in safeguarding people other than the operator since the results can be used to set boundaries for access by other people when the fields extend beyond the operator area. Whatever method is used for RF process machines, proper spacing and positioning of machines will reduce the possibility of other people being irradiated. Similarly, the avoidance of RF reflective materials near machines will reduce the possibility of field enhancements near the operator.

For machines operating at frequencies of less than 100 MHz, induced body currents are, logically, likely to become a significant factor in the control of operator exposure. The work of Gandhi[18] in connection with the measurement of induced body currents from people standing in a plane wave field and from people working with an RF sealer machine suggests that currents in the bodies of such machine operators could give significant ankle SARs in those cases reported where electric field values of 300 to 2700 Vm^{-1} and magnetic fields of 0.15 to 6.5 Am^{-1} were present.

The currents induced by the RF sealers were, however, appreciably less than those produced in the standing human beings experimentally exposed to plane wave fields.

Summarising the survey work likely to be involved with RF machines:

1 Measurements at the lower frequencies should involve separate measurement of the electric and magnetic fields as stated earlier under the 'RF surveys' heading. Measurements may eventually include limb currents and contact currents.

2 Most, if not all machines will need leakage tests applied to the RF source container or cabinet. For flow-line processing machines, leakage testing will extend to the structure around the RF applicator and the apertures where products enter and leave the applicator zone. X-ray tests, if applicable, should also be done.

3 Measurements relating to the exposure of operators feeding machines will be necessary where the operator is close to the applicator e.g. manual feeding. The worst case product, from the point of view of power applied, size, RF scattering potential, etc., should be used for measurements. Where the product is fed to the applicator automatically there may not be any exposure of the operator but access to the apertures referred to in (2) above may need to be controlled, both for the protection of the operator and anyone else who might approach the machine.

4 Control of flammable substances in the vicinity of the machine may need investigation.

Part 2 RF exposure surveys

RF radiation exposure measurements

General

It will have been noted from Chapter 8 that the practical range of RF exposure measurements can involve great variety and complexity, from the safety aspects of a single source to that of a complex site with twenty or thirty sources.

It is therefore only possible to illustrate practical exposure measurements by breaking the topic down into common elements. It is assumed that some approximate calculations have been carried out before surveys are started. This may be done by the methods of Chapter 5 or by any other accepted method.

When carrying out exposure measurements, the instrument probe should not be held closer to a source or a reflecting object than 20 cm. When working on stationary beams or in any situation where, for the purposes of the survey the potential hazard areas have been changed from the normal situation, adequate signs should be displayed to prohibit access. Otherwise people not familiar with the changed situation caused by surveying may inadvertently walk into a hazardous field.

Surveys of microwave beams

One common element of many surveys will be one or more microwave beams whether fixed as for some communications systems or moving as for radar equipment. The general technique used for moving beam systems is to stop the movement and align the beam in a direction which is convenient for the survey, but which does not cause any hazards. Fixed beams are, of course, normally surveyed in their working alignments. Tactical systems are a possible exception as they are usually easy to move, if required.

It should be noted that whilst rendering moving antennas stationary is the method used by many people when surveying, including the author, some people do use methods specifically designed for the assessment of rotating beams. These are described later.

Aims

Moving beams generally demand the most survey work due to their ability to irradiate people over their azimuth and elevation scans. The general objectives of such surveys include:

1 Assessment of the beam aligned at one or more elevations according to the nature of the site.
2 Assessment of the beam aligned to one or more representative elevated work platforms (where such elevated working is involved), or mast climbing paths, in order to determine the RF radiation levels present.
3 Making any representative measurements necessary at distances along the beam axis which correspond to the distance of flammable substances and EEDs known to exist on site, or at the proposed location of such new facilities.

   Obviously the beam cannot be pointed at a flammable or explosive substances store whilst measuring unless they are empty. It is therefore usual to point the beam in a safe direction and measure at a distance from the antenna corresponding to the distance of the store from the antenna. When extrapolating from a beam measurement on open ground to a flammable substances or explosives store, it would be wise to allow at least 6 dB for field enhancements due to reflections, which might occur at the store.
4 Assessing the safety compatibility of the operating antenna system with any other existing antenna systems – firing RF into another antenna of similar frequency could blow up receiver microwave diodes if there is any receiving equipment connected to it!

General assessment of beams

Figure 9.5 shows a plan view of a stationary microwave beam with a line representing the antenna axis. The beam elevation should be set to provide

Figure 9.5 *Plan view of an antenna aligned for a survey*

either the worst case or a specific elevation according to the needs of the survey. For antennas mounted close to the ground, the information will usually be obtained most easily with an elevation of 0°.

The author uses plastic traffic cones to provide markers, spacing them out as shown in the diagram, with the aid of a measuring tape. The intervals may be chosen for convenience according to the nature of the system. A spacing of 25 or 50 metres might prove satisfactory. Additional markers may be positioned where needed, for example to record measurements close in to the antenna either side of the beam axis line.

It will also be necessary to look at the beam in elevation as shown in Figure 9.6. Unless only man-clearance below the beam is sought then

Figure 9.6   *Elevation view of the Figure 9.5 arrangement*

measurements on axis at different heights will be needed, the choice depending on the survey requirements. For the purpose of providing an uncluttered diagram, three heights are shown in the figure. The surveyor will need a wooden (i.e. non-metallic) step ladder to take measurements. Checking heights above ground is usually done with a weighted non-metallic measuring tape. Heights which can be reached from the ground can be judged from a knowledge of the surveyor's own height or by using pre-marked survey poles.

There are some important preliminaries to be observed before starting measurements, since to walk into a beam without a careful check of likely radiation levels may put the surveyor and the instrument at risk. The initial steps to be taken are:

1 Check the calculations carefully. Ensure that the survey will not hazard flammable substances or EEDs stored on site, due to unsuitable choice of beam alignment. Also ensure that, if the equipment being surveyed is a pulsed equipment, the peak pulse power density is not likely to exceed any limits stated in the standard in use.

   If the limits are exceeded, the locations concerned will, by definition, be places where people must be excluded. (Multiply mean power density calculations by the reciprocal of the duty factor to assess peak power densities; repeat later with the measured values, if significantly different.)

2 Set the instrument to the highest range (greatest power density range) and ensure that the full scale power density value is adequate according to the calculated values.

3 If the system is a radar or other pulsed equipment, ensure that the measuring instrument is suitable for pulsed measurement and that the pulse duty factor (Chapter 8) is not such as to endanger the instrument.

4 If the equipment is a very high power one, check whether it can be run initially on reduced power. If the power attenuation factor is known, e.g. reduced to half power, the measured values can be scaled linearly. A cross check can be made on the attenuation factor by taking a measurement on axis at a safe place and then increasing to full power and checking the reading again, having left the instrument undisturbed in the same place.

5 If the instrument has a 'maximum hold' facility (Chapter 6) then this should be switched on. In starting the survey at a reasonable distance from the antenna, a slow walk across the beam with the 'maximum hold' facility switched on will record in the meter memory the highest power density met. This can be repeated at different heights and then closer to the radar until the relationship between measurements and calculations at ground level is established.

In comparing calculations and measurements, due allowance must be made for the possibility of enhancements from local causes such as metallic objects and structures, if these cannot be avoided. After making the required measurements at points previously defined on the general lines of Figures 9.5 and 9.6, the results should be recorded. The measurements may include the assessment of general safe distances for people and for flammable substances and EEDs.

In large or critical surveys it is usually desirable to duplicate the measurements with the aid of a second surveyor thus giving increased confidence in results. It is also important to check that the transmitter power output has not changed from time to time, either directly or by means of a portable radio contact. Radars can be monitored by listening for the p.r.f. on any small personal medium or long wave receiver to ensure that they are still working.

People are sometimes puzzled by strange readings near the ground round the feet although the main beam is above the ground at that place. The levels are usually low and can involve patterns of peaks and troughs corresponding to the wavelength in use. They usually result from reinforcing rods in concrete hard standing, roads and runways and are normally of no consequence except when significant energy is impinging on the ground e.g. with beams with negative elevations.

Measurements may also be needed around and at the rear of the antenna. All measurements can be recorded on test record sheets, for example with reference numbers allocated to each cone and alphabetical letters allocated to each height above ground, thus providing a logical basis for transcribing the data later on to copies of diagrams like Figures 9.5 and 9.6.

In some cases it may be preferable to put the test results in directly onto copies of drawings like those mentioned above, so as to get a visual impression of what is happening whilst actually working on the survey. This may be needed in order to determine what further survey measurements are to be made.

If it is necessary, the whole measurement sequence can be repeated at other beam elevations, taking care to watch for ground enhancements with negative elevations.

Beamwidth checks

It is often useful to do a rough check of the azimuth beamwidth using two cones as markers. This should be done in the far field wherever possible. It should be done at a height which approximates to the beam axis. A piece of wood of suitable height or a survey pole can be useful as a measuring stick to keep the instrument probe at constant height above ground.

Starting on the beam axis, move out across the beam until the meter reading drops to a half of the value on axis, keeping the probe at the same height all the time. Mark the half power point with a cone. Move back to the beam axis, keeping the meter height as before. Move out to the opposite side of the beam to the half power point and mark with a cone. The distance between the cones will be an approximation of the 3 dB bandwidth and can be checked against calculations.

In Figure 9.7, if the distance between the two cones is B metres and the distance on the beam axis to the antenna is D metres, the half beamwidth angle $\theta°/2$ is:

$$\theta°/2 = \tan^{-1} (\tfrac{1}{2}B/D) \text{ and } \theta° = 2 \times (\tan^{-1} (\tfrac{1}{2}B/D))$$

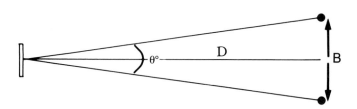

Figure 9.7 *Beamwidth approximation by measurement*

This method can also be utilised for lining up cones along the beam axis. The axis should be half way between the two cones. In practice, it does become automatic since, as measurements are done at each distance, the beam must be found and the highest reading at the given height and distance taken. Cones are then pulled into line.

This does not need to be be done with precision, but just accurately enough to find the beam. Surveyors have been known to miss the beam at the further distances from the antenna due to misalignment of the markers, with the consequent risk of missing significant RF levels or at least recording incorrect information. Hence it is important to find the beam at every measuring point.

Appraising the effect of beam elevation changes

When calculations have been done on charts like those in Chapter 5, the resulting beam has an elevation of zero degrees.

If during a survey it is necessary to get a rough idea of what will happen at positive or negative elevations, this is quite easy. Figure 9.8 is an actual

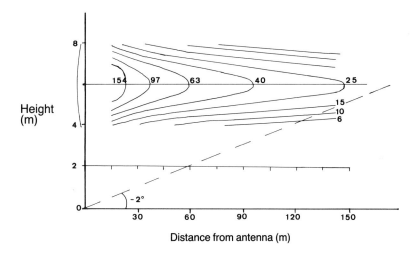

Figure 9.8 *Change of beam elevation on calculation diagrams*

calculation for the intermediate field of a radar height finder. It could, in practice, have been amended in the light of measurements already done if the calculations did not seem to be representative.

The chart 'X' axis does not, except by possible coincidence, represent the ground since this depends on the antenna mount and the effective height to the antenna centre.

This can be established from the scale of the 'Y' axis. In this example the height has been assumed to be 6 m and a new 'X' axis has been drawn below the existing one to represent the ground.

In order to assess the effect of the pattern being set to an elevation of −2°, a dashed line has been drawn tilting the ground up by this amount. If man-clearance is being examined, another line above and parallel to the dashed line could be drawn scaled, say, to 2.5 m. In many cases that height can be judged by inspection instead.

Examination of the diagram suggests that:

Somewhere between about 120 and 180 m distance from the antenna, the figures shown in the diagram should be considered to be increased by 6 dB due to the beam impinging on the ground with consequent enhancements.

It is likely, but not certain, that man-clearance for a stationary beam condition will be satisfactory up to about 50 or 60 m from the antenna.

This method of quick examination of 'what if' situations can be very useful. Angles can be pre-marked on the diagram and a transparent ruler used to act as a line. It is important to derive the angles by trigonometry and not with a protractor, due to the disparate scales.

Measurement of RF levels at elevated working places

The previous paragraphs dealt with the general assessment of the radiation levels in a beam under relatively ideal conditions near to the ground. This provides general information which can be extrapolated to other situations.

However, where people work in elevated positions either continuously or in the course of climbing a tower, it is necessary to align the beam, usually with the aid of an integral telescope, to the places under consideration.

Measurements are then made under representative conditions at the locations of interest. The levels to be expected will be roughly those found in the general beam assessment at the corresponding distance but with the possibility of up to 6 dB (power density × 4) enhancement due to metal structures at the elevated positions – more if a resonant antenna is the reflection object.

It will be possible to determine in advance whether it is necessary to do such measurements, from consideration of the results of the general beam assessment. If it is obvious that the power density on the beam axis at the distance from the source corresponding to the location being considered is low even if increased by 6 dB, then there will be no need to do a measurement there.

Sometimes with a fixed beam it is not possible to do a satisfactory assessment as described in the previous paragraph, either because the antenna is too high or because physical access cannot be obtained because there are ground obstructions which preclude the surveyor walking there and doing a direct beam assessment. In these cases it is necessary to measure at adjacent workplaces, one by one. Where the only problem is obstructed access close to the antenna, it may still be possible to do some measurements on the beam axis further from the source.

Measurements where the antenna is mounted high above the ground

Where the antenna is mounted high above the ground the methods outlined above are inappropriate. It will be necessary to operate the system in its fixed condition or if a moving antenna, in a worst case situation according to the objectives of the survey.

Measurements then have to be concentrated at the places where people are to work. Because the antenna is high it is likely that problems at the ground level will be reduced, but height finder radars and the like need some detailed investigation for the reasons mentioned earlier associated with Figure 9.8.

Use of hydraulic platforms for measurements

Sometimes there may be a special need to do measurements higher than is safe with step ladders. The use of hydraulic platforms (often referred to as 'cherry pickers') consisting of a long arm with a small cradle platform on the end such as is used for servicing street lamps, is about the only practicable method. There are measurement limitations due to field perturbation from the metal arm and cradle frame. The cradle itself is usually made of fibreglass or some other non-metal.

Figure 9.9 illustrates one occasion when such measurements were made on a prototype planar phased array radar. The measurements were plotted on

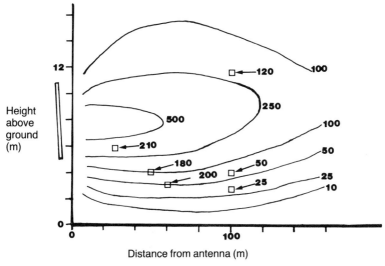

Figure 9.9 *Use of elevated platform for special measurements*

the original power density calculations. In the figure, the numbers not associated with little boxes are the calculated contour values. All values are in Wm^{-2}. The numbers associated with the square box markers are the measured values. It can be seen that at distances greater than 100 m the measurements though few, since this was not the area of interest, are comfortably close to the calculated values.

Closer to the antenna, reaching into the near field, there were some high values. Measurement was limited to the full scale value of $200\,Wm^{-2}$

available on the instrument. The operations are potentially hazardous and the method was only used by the author for special investigations close to the antenna in connection with the testing of a novel shielding material at high power densities.

Points to watch on surveys

When dealing with conventional microwave antennas the elevation setting scale of the feed horn should not be confused with the beam elevation. The latter needs to be established from an expression which gives the relationship between these two.

For surveillance radars having two separate back-to-back antennas, each needs to be surveyed separately. In assessing the rotational average power density, the results of the two calculations need to be summed.

Care should be taken to assess the effective radiated power correctly when two transmitters are used in diversity operation e.g. in radars. The treatment of their mean and peak power densities may be different due to the practice of staggering the pulse timings.

Where a telescope is fitted to an antenna e.g. at an aperture in a dish, this should be checked. At least one case has occurred where the telescope used in this way proved to be a waveguide at the frequency used!

Where sector blanking is in use, the limits should be checked.

Field measurements for personnel access

One of the most common survey objectives is to establish man-clearance, that is to say the ability to walk about at ground level in safety close to operating systems such as a radar or a fixed beam system.

Taking 2.5 m as the highest that an average person can correctly hold and manipulate an instrument and probe, then man-clearance requires the investigation of a volume constituted by the area in which access is sought times the 2.5 m height. In principle, this involves measuring all over the area at a number of heights up to 2.5 m.

With microwave frequencies, where the wavelength is short compared with the human body, there can be considerable differences in field across the section of a human being due to reflections from conductive objects and the ground. Hence the need to measure at a number of heights and spatially average the measurements.

The IEEE C95.1–1991 standard specifies that in such cases, measurements should be made at 20 cm vertical intervals up to 2 m and this would need to cover a width of about 2 m to represent a place where a person is to stand. For those who use 2.5 m for clearance it would seem appropriate to use 25 cm height increments in these circumstances.

This method would be very satisfactory for specific work places in small areas. Covering a large area on this basis would be an insuperable task and it would be more practicable to measure across the whole area at two or three heights and then inspect the data. It may be found by inspection that only a few places need the more thorough method since the levels elsewhere are low.

When doing man-clearance measurements it is important to remember that when beams are horizontal i.e. zero elevation, the measurements at ground level may be low for some distance out and then increase, as can be seen from the contours on the charts used in Chapter 5.

This obviously depends on the effective height of the antenna above ground as high mounts reduce the problem compared with low mounts. If the beam elevation is negative the increase in levels further out from the antenna may additionally be enhanced due to reflections from the ground. This can typically be 4 to 6 dB and more in those cases where reflection is from a resonant object e.g. spare antenna stored on the ground.

Figure 9.10 shows actual measurements on a 4 m dish antenna with 1 kW CW at about 4 GHz. The antenna centre was only 4.5 m above ground and

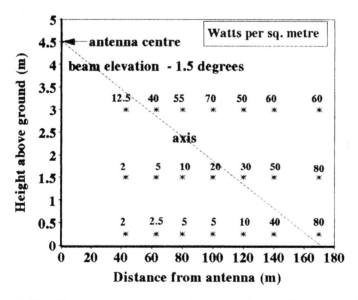

Figure 9.10 *Measurements on a microwave beam tilted towards the ground*

the elevation was −1.5°. In the diagram, the antenna beam axis is shown as the dotted line. At first sight the negative elevation may look much greater than 1.5° but this can be seen to result from the disparate X and Y axis scaling. This particular measurement was not done for man-clearance as the access to the beam area was prohibited, so that no measurements were made at 2.5 m height.

It is reproduced here to show the enhancement which can occur from the ground in a real case – in this case between 140 and 180 metres from the antenna. In the first 80 metres the levels are quite low up to 1.5 m height and inspection of the measurement data suggests that this might also be the case to 2.5 m.

Near field measurements

Because safety management involves the control of people who may work close to antenna systems, it is often necessary to measure in the near field. The measurement of power density, as discussed in Chapter 6, is not done directly but via either electric or magnetic field sensors, the instrument calibration being done on a plane wave basis. This means that technically, the power density measurements are not accurate in the near field.

However the quantity actually measured (Vm^{-1} or Am^{-1}) is correct and can be recovered by doing the reverse calculation from power density to the appropriate field unit. The generally accepted method is thus to measure the electric and magnetic field components and compare each with the permitted limits. Up to 300 MHz, this can be done and the near field – far field boundary from 30 MHz to 300 MHz is a relatively short distance anyway so that many VHF and UHF measurements may be in the far field.

Above 300 MHz it is not possible to measure the magnetic field as no instruments extend beyond that frequency. Consequently it is necessary to use wideband power density meters based on electric field sensors. The general practice is to measure power density and accept that the instrument will tend to give values higher than the real power densities in the near field.

This is due to the fact that the time phase relationship between the field components will be different to the normal in-phase plane wave condition.

There is no value in extracting the electric field value since if electric field limits are given for these frequencies, they will also be related to the power density limits on a plane wave basis.

It will be necessary under these conditions to use spatial averaging across an area representing the cross-sectional area of a human body to take into account changes of power density across the exposed person in near field conditions.

Characterising beam systems

It was noted in Chapter 8 that some equipments can be generally characterised in such a way that the safety provisions can be defined with some degree of confidence and incorporated in the equipment handbook.

This will usually only apply to the manufacturer but may occasionally be done by a major user.

It will mainly be the case where the nature of the equipment is such that siting will normally be clear of all buildings and people so that there is little need to change the safety provisions on that account.

An example of this was a field-portable communication system consisting of a dish antenna on a low mobile mount driven by 1 kW RF. This is used in the field with little chance of any buildings round it because of the need for a clear signal path.

The equipment was aligned on a flat surface about 1 km long and measurements carried out as illustrated in the previous paragraph, taking in a set of measurements for every half degree of positive and negative elevation in the elevation range. The results were studied and some contingency made for the practice of fine alignment of the dish in azimuth and elevation to secure maximum signal strength.

The result was the definition of a prohibited area in the form shown in Figure 9.11. This is a rectangle having a fixed dimension with the length varied according to a table which reflected the safety levels used by different customers at the time. A general height prohibition was also added although this was a formality as no one was likely to use a crane or anything similar over the area.

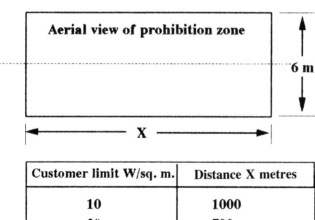

Customer limit W/sq. m.	Distance X metres
10	1000
20	700
50	300

Figure 9.11   *Mobile communications equipment–definition of hazard zone*

Measurement of rotating beams by special methods

It was mentioned previously that some people like to use measurement methods which do not require rotating antennas to be rendered stationary. Three methods have been put forward by the Narda company.

The second method, and the third which is a variant of the second method, became available later than the first one and may, possibly, be considered as superseding it. It would seem to be desirable to check the methods against a stationary measurement, at least when first using them.

The measured values resulting from these methods are the rotationally-averaged power densities at the place of measurement.

Method 1 Instrument time constant method

This method requires the time constant of the Narda instrument to be specifically calibrated. The Narda notes on this topic were written for the series 8600 radiation monitors[51].

The 8600A time constant of probe and circuitry is not constant, varying with the illumination time. The time constant of the type 8600B probes is not subject to this sort of variation and can be calibrated by the manufacturer on request.

For the type 8600A probes this necessitates the selection of an effective time constant from a table which relates illumination time and time constant. For the purpose of the example here a time constant is assumed on the basis of the original Narda paper.

The aim is to provide a measurement at a point illuminated by a rotating beam which will correspond to the measurement which would be obtained if the beam were stationary and aligned to the measurement point. The method uses the instrument in the 'maximum hold' mode which gives a reading corresponding to the highest value of power density seen by the instrument during illumination by the beam.

For short illuminations, several illuminations (rotations) of the beam may be required to reach the full reading.

The illumination time of the beam at the measurement point can be calculated from the beamwidth at that point and the rotation rate of the antenna.

Example:

Assume that the 3 dB beamwidth is 1.3° and the rotation rate 12 rpm.

For a rotation rate of 12 rpm, the time of 1 revolution (360°) is 5 seconds. The illumination time t is therefore:

$t = (1.3°/360°) \times 5$ seconds $= 0.018$ s.
assume the corresponding time constant $T = 0.291$

The ratio of indicated to actual power density (K) is:

$$K = 1 - e^{-t/T}$$

now $t/T = 0.018/0.291 = 0.062$
and $K = 1 - e^{-0.062} = 0.06012$

The 'maximum hold' reading must be multiplied by $1/K = 16.63$ to give the equivalent 'stationary beam' reading at the measurement point.

Note that the method is not limited to a 360° scan and may be some lesser figure. However the scan must be cyclic.

Method 2 Measurement of fast-scanning radars using a Narda radiation averaging module unit type 8696 [52]

The method involves three specific activities which take place after the instrument has been given adequate time to warm up.

1 Checking the instrument zero at the start and finish so that it can be averaged.
2 Running the instrument for the survey and obtaining the survey result.
3 Correcting the survey result for the average zero error.

With the instrument out of the influence of any field e.g. in the carrying case, it is set up on the highest range which will not allow the meter to become overloaded and the meter zero set. The mode switch is set to 'spatial' and the start–stop button pressed. After five seconds the button is pressed again and the number indicated on the display is recorded as the first zero.

The beam survey is then carried out as a time-averaged measurement and the survey average is read from the meter as a percentage of full scale.

The zero check is then repeated and the reading recorded as the second zero.

If the survey meter reading is X% of full scale and the before and after zero errors are Z_0 and Z_1 respectively, then the zero correction is made by averaging the sum of Z_0 and Z_1 and then deducting it from X%:

Active survey corrected average $= X - (Z_0 + Z_1)/2\%$

The power density is then obtained by reference to the full scale value of the range used.

An example given by the Narda Company used the following data:

First zero $= 0.1$
Second zero $= 1.5$
Survey six-minute average from meter 15.8% of f.s.d.

Zero average $= (0.1 + 1.5)/2 = 0.8$
Corrected survey average $= (15.8 - 0.8)\%$ of f.s.d. $= 15\%$ of f.s.d.

Method 3 Measurement of slow-scanning radars using a Narda radiation averaging module unit type 8696

This method is very similar to the previous one except for the additional use of a stop watch and the method of carrying out the actual survey using the 'pause' control and 'time' recall facilities of the module.

The activities are as follows:

1 Checking the instrument zero at the start and finish so that it can be averaged (exactly as method 2).
2 Recording the elapsed time duration of the survey by means of a stop watch or wrist watch.
3 Running the instrument for the survey and obtaining the survey result and the time duration of the survey, which will differ from the elapsed time record due to the method of use of the instrument.

The instrument is set up in the same way as before out of any field and the initial zero error is recorded. The start button is pressed and the stop watch started. As the beam passes the survey point and the meter returns to zero the pause button is pressed once.

As the beam approaches the measurement point again, the pause button is used to switch the meter on and then off again after it has passed. This sequence is repeated for at least 10 passes of the beam. 'Stop' is then pressed and the stop watch also is stopped.

The survey average is read from the meter as a percentage of full scale and the survey time is also read from the meter by switching to 'time'.

The second zero is then obtained and the correction calculated as in the previous method.

The ratio of survey time to elapsed time (meter time divided by elapsed time) will be a number less than one due to the fact that the instrument was only switched on for the period of time when the beam illuminated the meter probe. The corrected survey average above is multiplied by this fraction to arrive at a final survey value in percentage of full scale and hence in power density.

An example given by the Narda Company used the following data:

First zero = 0.3
Second zero = 3.2
Survey average from meter 28.9% of f.s.d.

Survey time from meter 32.6 s.
Elapsed time from watch 180 s.

Zero correction
(3.2 + 0.3)/2 = 1.75

Average for active survey (28.9 − 1.75)% = 27.15%;

Time correction
Ratio of survey time to elapsed time = 32.6/180 = 0.181;

Final average = 27.15% × 0.181 = 4.91% of full scale.

This can be changed into power density by reference to the range full scale value.

Special cases of movable beams

Conventional moving beam antennas

As discussed in Chapter 5, beams which do not rotate or otherwise move on a cyclic basis and which may, therefore, dwell in a particular orientation, cannot be treated on a 'rotating beam' basis if the result could be to irradiate a person at levels exceeding the permitted limit. This can apply to height finder radars, tracking radars and to any irregularly moving system which may irradiate people in excess of the relevant standard by either dwelling in their direction or by making a large number of passes in their direction in a short period of time.

Obviously, if a beam is virtually stationary, tracking a target on a bearing which corresponds to that of the people being exposed, the dwell time might be long if the period of interest in the target is significant and the target track is a directly approaching or receding one. Whether this can happen will depend on the relative height and position of the people liable to exposure.

The requirement to use the stationary measured power densities implicit in the possible continuous irradiation of people is generally referred to as using 'stationary beam criteria'.

Electronically-switched beam systems

Another category of moving beam is the spaced array system where transmit beams can be electronically switched without any motion being involved. Whilst mechanical motion can easily be detected and the failure indication used to switch off the RF, it is not necessarily possible to provide such a system to determine when something has caused a beam to switch incorrectly and expose individuals.

In the case of spaced array antennas where there is no visible motion, it is usually considered necessary to use the worst possible case to produce safety criteria based on the assumption that the beam has failed to that worst case situation unless there is an in-built failure detection system.

For these modern systems the survey methods need to be determined according to the nature of the system and the site. Often where people and equipment can be adequately separated a simple safety boundary and hence a simple survey may be possible.

Spatial and time averaging methods

Spatial averaging

Spatial averaging is used to average measurements over areas and volumes. An example has already been given in respect of averaging over an area, the height and width of a person in a field. Repeating this process over a ground area and averaging the sum of the averages of each set of measurements effectively averages a volume. This is a fairly tedious task, though made easier by the use of an instrument averaging facility.

In practice, when the levels being measured are well below the permitted level e.g. when checking rooms which do not contain RF sources, the measurements might be simplified by limiting them to perhaps three vertical intervals rather than the 20 cm vertical intervals mentioned earlier.

The manual method is usually to survey the area slowly at the first height interval, recording the highest field of any reading found and then repeating at the the other two intervals. Where significant readings are found, a vertical sweep from ground to the highest level specified is done to see whether there are any readings higher than the three being recorded. If the three readings are the highest ones anywhere and well below the relevant permitted limit, the three can be averaged to give a worst case spatial average.

This conservative approach obviates the need for recording many readings for each sweep and doing arithmetic averaging operations to no good purpose except to end up with a lower overall average. However, if the readings found are close to or over the limits, it will be necessary to do formal averaging on the tedious basis mentioned at the beginning of this paragraph.

With the ability to use the facilities of an instrument, spatial averaging can be done at the press of a button and most of the data recording is unnecessary. From the manual example above, the amount of work saved in recording readings when full averaging is done is obviously great. In using spatial averaging it is obviously necessary to move sufficiently slowly to allow the instrument to read the value at each place properly otherwise incorrect averages may result. Vertical and horizontal sweeps can be done easily.

Time averaging

The limits for field quantities given in most standards are, by definition, the average values over a time period which is specified. For occupational purposes the averaging time is normally one tenth of an hour (6 minutes) at least up to 15 GHz. There are three ways of using time averaging:

1 Where the radiation field varies unpredictably with time. Averaging this by manual time averaging is tedious and rarely, if ever done.
2 The special methods of measuring rotating microwave beams earlier in this chapter which provide for rotational averaging directly and are time averaging methods.
3 Where a short exposure is involved. For example, consider the case where it is occasionally necessary to walk straight through an area where the measured power density is, say, twice the permitted limit. If the person was able to walk through in less than 3 minutes, the time averaged power density would be acceptable providing there was no exposure in the next three minutes.

Time-averaging involves recording the meter reading continuously or at frequent intervals and at the end of the time period presenting the time-averaged value. The methods which can be used are:

(a) Recording the output of the meter from the meter recorder output terminals on a chart recorder which has an accurate time axis. This gives a picture of what is happening but does not offer averaging unless the recorder is particularly sophisticated. Sometimes the recording can be averaged by visual inspection, but otherwise it may have to be done by tedious analysis. An alternative is to pass a data output from the measuring instrument to computer storage and perform averaging afterwards. This would usually need a fibre optics interface.
(b) Using an instrument which has time averaging facilities included. Instruments offering time averaging facilities are able to do averaging automatically. They offer one or more pre-set averaging times which can be selected. It is then only necessary to press a start button and the instrument will stop and display the average at the end of the time period.

With an automatic measurement system it has to be accepted that large readings caused by interference, which would be noted when doing normal manual measurement will get averaged with other readings. It may, therefore be desirable to run the time averaging several times if such interference is known to be prevalent.

Ground communications systems

There is an enormous diversity of equipment used for communications purposes from microwave to the lowest frequencies used for communications. Microwave communication equipment can be treated as detailed in the foregoing text.

Apart from microwave equipment, UHF, VHF, HF and MF equipment may be involved in communications. A wide range of modulation methods may be used from CW keying (morse code) through all the other possible forms. The transmission of speech and digital data are two common functions of communication systems.

MF and HF wire and mast antenna systems

Fixed MF and HF systems may use large wire antenna systems occupying many acres of land. Lower frequency systems may use a metal tower as the radiating element. Feeders may take various forms from coaxial cable to open-wire transmission lines.

The usual practice is to determine safety boundaries for antennas and prohibit access to those areas. Since HF systems may change antennas during any twenty-four hour period for propagation reasons, the overall pattern for a site can be complex and surveys need to explore all the antenna combinations used.

Since only technical staff will normally be used within the antenna area of the site, control of access should be exercised via a 'permit to work' system plus the declaration of any necessary prohibited areas.

The complications that arise include:

The safety of riggers who may have to climb towers to change antennas or to maintain them. There is a possibility of burns both when climbing and when moving around the site due to the electromagnetic fields present and the distribution changes which take place as antennas come in and out of use. Major broadcasting organisations have specific procedures for climbing to cover both of these aspects, particularly for the routine maintenance work such as painting masts, etc.

At MF and HF frequencies the magnitude of electric field strengths needed to ignite flammable vapours or fire EEDs are considerably less than for VHF and UHF. Apart from any such substances stored, it may be necessary to control the use of such materials in maintenance, since this activity can result in the substances deliberately being brought close to antennas.

To emphasise the importance of this aspect at these frequencies when carrying out surveys, Figure 9.12 is included. This was created by calculating values for 1 to 30 MHz based on threshold data from BS 6656:1991 [31] for gas group 2c. Two sizes of pick-up loop size (10 m and 65 m) have been assumed.

The figure shows the threshold electric field strengths for the conditions mentioned, higher values giving rise to a risk of ignition. It can be seen that the thresholds are surprisingly low at HF. Although other standards may apply to many readers, the order of the levels should be similar since they derive from practical research.

The importance of this is also to illustrate that the ignition of flammable vapours can take place at levels which are well below the limits for human

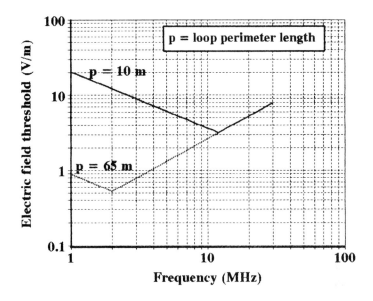

Figure 9.12 *Field strength safety limits for the 1 to 30 MHz band calculated from BS 6656:1991 for flammable vapours, gas group 2c*

exposure. Some people assume that what is safe for people is safe for flammable substances.

Radiation levels are generally high very close to open wire feeders and external antenna matching units, as they are when close to arrays. Time averaging is generally used to safeguard the surveyor by permitting short inspections of suspect feeders. This means a short exposure followed by withdrawal from the field to meet the six minute averaging provision.

As noted earlier, frequencies below 30 MHz (or 300 MHz according to which standard is used) are not assessed in terms of power density but rather, the electric and magnetic fields are separately measured and compared with permitted limits. This obviously applies to MF and HF frequencies. At very low frequencies some of the measurement instrument problems listed in Chapter 8 should be taken into account.

The aim of a survey of a large antenna field will depend on the nature of the site, the interaction with the public and the degree of operational technical access to the antenna field required. It is therefore impossible to generalise as to method. However, the likely aims will be:

To determine the permitted limits of access and therefore the boundaries of areas where access is to be prohibited. Note: It is important to allow for multiple irradiations, where applicable, in all safety assessments (Chapter 5).

To determine the safety of climbing and other technical access requirements and the procedures by which these may be controlled.

Verifying that the public are not subject to exposure in excess of permitted levels and also that they cannot incur any burns in contacting metal objects in the public domain.

There may also be a need to look at certain EMC aspects such as the safety of motor vehicles on roads close to the antenna in respect of any vehicle electronic control systems.

Checking hazards to flammable substances on and off site; similarly, EEDs need to be considered, if relevant.

The method of surveying will depend on which of these factors are applicable. It will generally start with a rough assessment of the antenna patterns and the magnitude of the fields at convenient points. This will then give a starting point for deciding further measurements.

From the surveying point of view, one simplification when possible, is to make as large a prohibited area round complex systems as practicable, rather than do an enormous amount of work to make it small. If no one needs to go into such an area, or it can be controlled for infrequent access by a 'permit to work', then there may be no good reason for increasing risks by arranging the smallest possible area.

Surveys involving a large area of land are very tedious and expensive. The problems can be simplified to some extent by formalising human access on site to defined pathways which are suitably marked as opposed to allowing people to choose their own routes. For administrative purposes the other areas should be deemed as 'not surveyed' and classified as prohibited areas.

This can allow the survey to be concentrated on the pathways. The difference in the amount of work can be appreciated from consideration of Figure 8.6 in Chapter 8, which itself is only a small site. Some sites may have twenty or thirty transmitters and many antennas.

UHF, VHF and HF whip and rod antennas for ground use

These may be simple rod antennas or arrays of rods. Most simple rod and whip type antennas are used on mobile installations where they form part of the vehicle installation. Arrays may be used in fixed applications or in portable systems. Portable systems, in contrast to mobile installations, may carry masts and more directive antennas for erection at the place of use.

The rough calculations for such antennas can be done by the methods given in Chapter 5 which contains a typical table. A knowledge of the power into the antenna and the antenna gain is all that is needed. Where such antennas are used in ground installations, the survey methods are quite simple and give few safety problems if the antenna is sited properly so that where necessary a small access prohibition around them can be specified.

Where such systems are badly sited by being placed amongst people or close to anything which could be hazarded then, often, moving the antenna is the answer. For HF, particular care is needed to keep unnecessary metal objects out of the field and hence prevent burns.

For most simple antenna systems which do not occupy a lot of ground area, simple prohibited access boundaries will be the best method of ensuring safety where the fields present necessitate it. In the occupational situation where only adults could access the area, often ropes or plastic barriers and suitable signs are used.

Tower mounting is often used for communication antennas. A useful example is for air traffic control (ATC). The antennas are usually VHF and UHF dipoles with a gain of a few decibels and there may be quite a large number fitted on each tower (see Figure 9.13). Safety assessments on towers are difficult due to operational conditions.

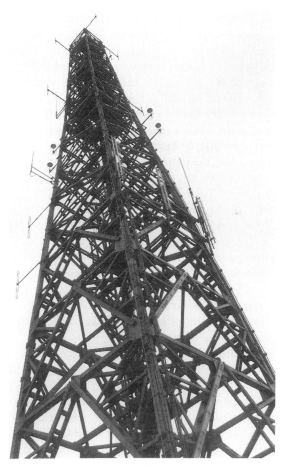

Figure 9.13 *ATC tower with VHF and UHF antennas (Courtesy of the Civil Aviation Authority)*

Looking at it from a practical point of view, there are two main activities involving people climbing towers – maintenance on the tower and maintenance on tower mounted equipment. Climbing central ladders will usually be safe since the antennas are normally mounted about 1.5 to 2 metres out from the tower so that the distance from the centre of the sort of tower often used for ATC will be perhaps 3.5 metres or more except at the top of the tower.

Safety checks can be done at a time when radio traffic is known to be highest. Tests of this sort with groups of antennas on the mast each fed by 40 to 50 W transmitters and operating as required by the aircraft traffic, produce quite low readings or none at all on the centre of the ladder.

More powerful transmitters may give different results.

Statistically, the surveying climb may need to be done a number of times and last for long enough each time to get a representative reading over a busy traffic period. The general task of climbing towers is often complicated by other factors such as potential irradiation by radar beams and by the increasing tendency for all sorts of antennas to be added to existing masts, including equipment for people not directly associated with the site e.g. radio telephone equipment and other local communication equipment, microwave dishes and the like, owned by other operators. It becomes very difficult to calculate the likely exposures since the times at which individual systems transmit is seemingly random.

On a time-averaged basis there may still not be a problem with a steady climb but when work is to be done on the tower, specific arrangements may be necessary to deal with any excessive exposure from particular sources. It may be necessary to do some representative measurements at such work positions. If an antenna is to be repaired or replaced, this might also mean checking that it will not give rise to burns due to being driven parasitically by one of the other antennas!

Not all communication towers are generously proportioned. Many towers are very slim and may have to be climbed on the outside, thus bringing the climber much closer to antennas. Climbing and working on such towers should be controlled on the basis of survey results so that overexposure can be avoided.

Mobile vehicle-mounted systems, especially HF systems, pose particular problems. Experience in the survey of such systems has shown up a number of factors which need careful attention during surveys:

HF antennas driven with significant power and mounted on the side of vehicles can present hazards to those standing near the antenna and can be a serious worry in the public domain unless access can be restricted. The hazards include burns and exposure to fields which exceed the limits of the relevant standard. Careful and systematic measurements are needed over the frequency band. This is one activity where the new contact current meters now coming onto the market could be useful in checking vehicle structures.

VHF and UHF are likely to provide less potential for burns and shock but need to be checked thoroughly for odd resonance effects with structures and objects. In almost any mobile system which is connected to remote lines, it is desirable to avoid contact with the lines at the vehicle end when operating, since they are inclined to collect RF energy directly.

Satellite ground stations

Fixed satellite ground stations with large and threatening-looking large dishes do sometimes cause public concern, mostly quite unjustifiably.

The beam is very narrow and directed at a satellite, the elevation of which depends on the location of the station and the satellite concerned. Antenna systems are normally fitted with elevation limit switches which stop transmission or antenna motion when any attempt is made to drive the antenna below specified elevations.

Surveys will normally be concerned with local radiation applicable to station personnel, checking the correct operation of limit switches and other local control systems. Practical surveys show that, for most sites, checks outside the site will largely be for psychological purposes, since significant radiation at ground level (i.e. levels near to the limits of the relevant standard) is normally not experienced.

Portable satellite systems may need treating differently depending on the circumstances of location and operation and on the power and system characteristics. For example, a portable system deployed in the public domain may need to be characterised as discussed earlier to define a safety zone around it, with the provision of any necessary safety instructions to be followed by the operator.

Broadcasting on VHF and UHF

High power broadcasting transmitters for television and VHF usually have antenna systems on high towers often located in rural situations. In such circumstances the power density on the ground in the public domain is low, the hazardous fields occurring around the actual antennas. The main problem which involves survey measurements is the need to climb towers for antenna work, tower maintenance and the maintenance of aircraft warning lights on towers.

Climbing is likely to involve near field exposure and hence the previously mentioned interest in trying to find some sort of 'true power density' measuring instrument, since the conservative readings of conventional instruments are not helpful. The problem is a difficult one, especially where towers are shared by different broadcasting organisations, so that tower climbing may involve radiation from antennas belonging to the other organisation. Typical antennas are described in Chapter 2.

The VHF antenna is often located below the UHF antenna and the radiation levels for a high power system can be typically 43 to 137 Vm^{-1} in the climbing space behind the VHF tiers with 'hot spots' exceeding 275 Vm^{-1}. Where it is necessary to climb through this, it is obvious that it may be necessary to switch off or switch to lower power (a presently used method).

Broadcast antennas situated on the top of buildings in highly populated areas may pose other problems. Survey methods therefore need to be tailored to the situations obtaining and the safety management methods will depend on many factors including station broadcasting times, whether maintenance can be done during close-down, whether power can be reduced, etc.

A useful paper on the calculation of power density from UHF and VHF broadcasting transmitters has been produced by Jokela of the Finnish Centre for radiation and nuclear safety[65].

Avionics and ship systems

Avionics and ship systems have in common the fact that they move about whilst operating powerful radar, communications and other equipment.

Avionics systems

The usual requirement for avionics systems which run in the proximity of people when taxying, taking off and landing is to determine safe distances for people and for other hazardous materials such as flammable substances and EEDs. When stationary with equipment operating, the same provisions apply plus any specific servicing safety provisions when a bay or an electronics pod is opened.

Safe distances can be established by calculations and confirmed by measurement. Contingencies need to be added to allow for enhancements from reflections.

Radiation arising from maintenance cannot be calculated and periodic monitoring is only a partial safety measure since defects may change the amount of radiation. Local in-situ monitoring is one current method using personal pocket monitors or a universal monitor which can be placed in the general area (see Chapter 6).

Ship systems

There is an increasing tendency to have more and more transmitting equipment on ships, especially warships. This means a corresponding number of antenna systems, some of which are well above deck but with others relatively close to walkways and personnel. The climbing of structures will need to be controlled to avoid unexpected hazards to people.

For fixed antenna systems, one method is to define safety distances after measurement by painting circles on the deck around antenna systems. The fields inside radio cabins using MF and HF equipment can be high round the feeders. Whether this significantly affects the operating personnel depends on the layout of the cabin. In some cases, open feeders, usually copper tubes, may need to be shielded. Some ship structures such as wires, walkway rails and the like may become burn hazards due to induced energy, particularly if of resonant dimensions.

Rotating microwave beams such as radars need surveys to investigate possible exposure of people. Controls are needed to prevent hazards when docking, especially those effects which might put flammable substances and EEDs in the dock area at risk. There can also be effects on dock side machinery electronics e.g. safety systems on cranes and other lifting gear. Similar problems can arise in respect of stores being loaded either in dock or at sea.

The extent of such risks depends on the nature of the vessel and the immediate environment, i.e., dock side, other ships, etc. Some operations can be safeguarded by determining safety distances for radars and other high power microwave beam systems and observing them. Other operations, particularly loading hazardous stores, might require the system to be closed down.

Handling survey data

The amount of survey data collected varies enormously according to the nature of the task, the number of sources involved and the physical arrangement of the equipment on the site. The data collected should be related to the measurement plan for the survey. Each planned measurement or series of measurements should have an objective and the results can be tabulated against the objectives.

For simplicity of presentation, leakage surveys and exposure surveys have been treated separately below, many surveys will involve doing both types of survey work.

Leakage surveys

For this type of survey, the situation is, in principle, quite simple. Leakages greater than the permitted limits, whether to a standard or against a contractual commitment need rectification. The important thing is to ensure that the remedial work is clearly defined. There must have been many cases where the wrong waveguide flange or other part was stripped down needlessly, because of some ambiguity in the instructions.

Whilst a lot of leakage data will have been recorded and can be included in the survey report, the reader will perhaps be best served by an easy summary table which only lists those leakages where rectification is needed. A complete table full of figures which are mainly acceptable distracts from the task in hand, which is that of dealing with the unacceptable.

Exposure surveys

It is convenient to examine human exposures and any other factors such as flammable vapour hazards separately.

Human exposures

1 Rotating or scanning microwave beams

Where it is permissible to do so, rotationally-averaged power densities can be calculated as in Chapter 5, for the places of interest.

2 Fixed antenna systems

The actual measured power densities or other field quantities will be used in these cases.

3 Multiple irradiations

The possibility of multiple irradiations will have become apparent during the survey and need to be allowed for using the expressions in Chapter 5 or those in any standard being used.

4 Man-clearance

This can be established from the measurements taken. It may be conditional (dependant on specified operating parameters) or unconditional. The maximum height used should be stated.

5 Prohibited zones round equipment

These should be determined from the measured radiation levels and the zone clearly defined. It is usually, though not necessarily, a circle. The area should be visually inspected if this has not already been done, to see whether there are any possible sources of enhancement which might have been overlooked. If not already detected and measured, this should be done. It will be likely that the result will be to increase the size of the prohibited zone or modify the shape of it.

6 Personnel working at elevated positions

The locations concerned should have been the subject of specific measurements and that should determine whether work exposure at each location is acceptable.

Climbing a tower to service equipment on the top and passing through a beam on the way up may be acceptable on a six minute average basis, although it may not be acceptable to stop and work at the place where the beam irradiates the tower. If this is the case, there should be appropriate markings on the tower.

Where any clearance to work is conditional, the conditions should be clearly stated. For example:

'Operation on platform X is permissible only when radar Y is operating at a positive elevation of 1.5° or greater'.

(Consolidates what was probably a verbal agreement with the operating authority on the minimum positive elevation of the radar. This should then result in a suitable instruction being issued to operating staff.)

'Tower Z may be climbed by authorised technicians to service equipment A at the top of the tower providing that climbing and descent is continuous through the small part of the tower which is marked with hazard warnings; for any other type of work no climbing is permitted when transmitter No.1 is operating, except with a specific permit to work.'

(An example of a case where one well defined activity is authorised and all others are to be dealt with by a permit to work.)

7 MF and HF antenna systems

When considering safety recommendations in situations involving wire antennas and arrays it is not only important to look at fields from antennas but also to include:

(a) Likelihood of burns being incurred from conductive objects.
(b) The fact that transmitter antenna arrays are also good receivers; a 'not in use' antenna may be energised from an 'in use' antenna and the terminals and structure of the array might give rise to shocks and burns.
(c) Any burn and shock hazards likely outside the site i.e. in the public domain.

8 Hazards to flammable vapours and EEDs

The key factors to remember are:

(a) Modulation should be taken into account when assessing field levels and the risk to flammable vapours.

(b) With pulse transmission, peak power density or peak electric field strength is the relevant parameter, not mean power density or mean electric field strength.

(c) For EEDs, the relevant parameter will depend on the thermal time constant of the igniter and guidance can best be obtained from expert sources including any standards dealing specifically with electro-explosives.

10
Designing to reduce radiation hazards

Introduction

When RF and X-ray surveys are carried out on sources of RF energy the survey findings will, from time to time, require some remedial action to be taken. Where such surveys are carried out on prototypes and pre-production models of new equipment designs, a body of knowledge about common design weaknesses is created and can be used by designers to guide future design work.

This chapter outlines the sort of information which can be used by designers and also that which can assist those concerned with marketing to avoid the problems which can arise with the siting and installation of radio transmitters and other sources of RF energy.

Experience shows that there can sometimes be a lack of knowledge on the part of the designer of the hazards of RF radiation and the methods used to reduce the hazard. This is particularly the case where a designer has just moved from very low power work where relatively little knowledge of RF radiation safety was needed, to the design of higher power equipment where the situation can be very different. A similar possibility occurs when a designer moves from low voltage equipment to high voltage equipment without much experience of X-ray radiation from the high voltages applied to high power electronic tubes.

The design process

The product design process is a complex one and is conditioned by marketing policies and objectives. Products designed for particular customers may be designed against individual requirements specifications, either written by the customer or produced in a generic form by some body or organisation having an interest in that type of product. Products designed for a general market have to take into account the views of those purchasers known to the producer and any lack of knowledge of the market may have adverse consequences. This is true of most aspects of a product e.g. performance, function, structure, cooling methods, and so on.

Often the wider implications such as radiation safety may be inadequately investigated so that customers find the product does not adequately meet their own requirements for the safety of their personnel. Also, the approach to radiation safety does not, in the case of most types of RF power source, lend itself to a simple 'safe product' concept. This arises from the nature of radiation since, to take the example of the RF transmitter, the aim is to radiate RF energy efficiently and as a result, it will rarely be safe to stand close to the antenna except with very low power systems.

Evidently, in the case of transmitters, process machines and similar RF power sources we can only talk about a safe installation where the concept of safety involves a mixture of safety provisions such as the control of unwanted radiation from cabinets, feeders and other items which may be accessed frequently by the operating staff and by the control of access to the point of intentional radiation.

In the case of fixed ground transmitters, this involves the exclusion of people from the proximity of an antenna, which may not be a single antenna but a choice from a number of antennas depending on the time of day and the required frequency. This may require barriers, management control and, in the worst cases, electrical interlocks.

It follows that the equipment user cannot look to the equipment designer for complete radiation safety where any hazard arising stems not from the intrinsic design but from the siting, use and management of the product. He can, however, expect guidance from the product producer about siting and use. Further, where transmitters and antenna systems are involved, the wanted radiation may impinge on adjacent people and installations outside the control of the owner of the transmitter.

Sometimes the product producer takes on a 'turnkey' project where he supplies a complete station or system and takes responsibility for all aspects of it. In these circumstances, it is essential that the site and the surrounding buildings and installations are taken into account in the station or system planning process.

Figure 10.1 identifies some of the questions which need to be considered both when advising the customer generally and when taking overall responsibility for the project.

1 Flammable vapours

Standards exist for the identification of risks involving flammable vapours which provide some guidance on the nature of the problem, outline how to approach the subject and undertake calculations to determine whether there

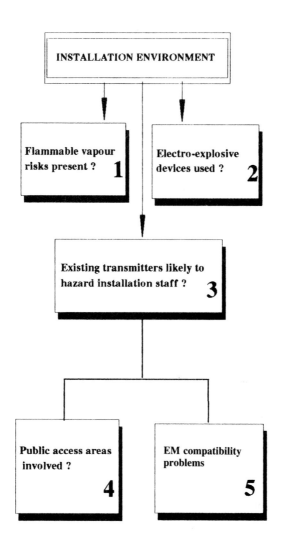

Figure 10.1 *Flow diagram – the installation environment*

is a hazard. A typical standard, British Standard BS 6656:1991[31] provides such information and also gives some tabulations of transmitter types and powers with the relevant safety distances, which can be used for an initial appraisal (see Chapter 5). In any specific case, the national standard concerned should be used.

Possible commercial and industrial installations which may need investigation include gas terminals and distribution systems, petroleum and similar installations for storage, processing and distribution and other processing plants involving flammable vapours.

2 Electro-explosive devices (EEDs)

This topic is conceptually similar to that above for flammable vapours but can be split into two parts – civil and military devices. Commercial detonators are covered in some standards including British Standard BS 6657:1991[32]. Commercial usage includes quarries, stores and explosives manufacturing plants.

Military explosive risks need to be evaluated in conjunction with military experts. Because of the operational handling and movement of stores containing explosives, the situation needs constant review.

3 Existing transmitters on site

When a supplier secures a contract to supply further equipment to an existing site e.g. an additional radar, a problem which often arises at the time of delivery of the equipment is the realisation that it has to be erected and installed where there is a significant RF radiation field from existing equipment. This can lead to considerable problems with installation personnel who generally find out about the problem when they notice sparks from the tools they are using! It is clearly important to establish a procedure for safe installation as part of the contractual negotiations.

4 Public access areas

In countries where public paths and access routes are established in law and mapped on national maps, it is necessary to ensure that these are recognised and antennas sited so as to avoid radiation fields in excess of the nationally-established safety limits.

Public access is not limited to paths and also includes residential occupation and the use of land for any purpose. In all cases, an essential requirement is the avoidance of the possibility of the public incurring shocks or burns from conductive objects in their possession (vehicles, bicycles, agricultural equipment, etc.) or from objects located where the public have lawful access such as scrap metal, metal clad buildings, caravans, parked vehicles and similar potential shock and burn sources.

5 Electromagnetic compatibility (EMC)

Unwanted interaction between RF radiation sources and other equipment is not always to be seen as a safety issue. The effects may lie between safety and economics. At the one extreme, EMC problems with an aircraft may, in the worst scenario, interfere with the aircraft control system and hazard the aircraft.

In lesser cases, communication may be jammed. At the other extreme, interference experienced by an administrative computer may merely be a nuisance with an economic rather than a safety consequence, i.e., extra computer time needed to do a task.

When surveying likely sites for new transmitter installations or for the additional installation of further equipment, it is virtually impossible to quantify the potential EMC implications for the local environment and it is necessary to err on the safe side when making judgements.

For new sites, the acquisition of extra land with which to separate the RF source and the local inhabitants, roads and other potential problem areas is one of the options, though this carries a cost penalty. On existing sites the degree of freedom may be limited and more care is needed with the siting of extra equipment.

Whilst it was noted above that EMC problems may often merely be in the nuisance category, this should not be treated too lightly. In the eyes of the public there are few electromagnetic 'sins' more unforgivable than interference with radio and television reception! On occasions, a portable television set may be found to be a useful piece of test equipment.

Planning design from the radiation safety viewpoint

Introduction

It is difficult to look at the radiation safety aspects of product design in isolation from the whole process of design and the object of this section is not to promote isolation but instead to identify those elements that should be included in the overall design plan. The actual solutions to specific radiation problems will not always be the obvious ones since designers have to trade-off any resulting effects on performance and cost.

Competent transmitter designers do not need telling how to do their jobs but generally do appreciate the sort of guidance on RF radiation safety which will reduce the likelihood of expensive rework to reduce excessive radiation. It is important to recognise that most transmitters have a long life so that, in a climate where safety limits in standards are constantly being tightened, design should be conservative in this respect, if later, costly, modifications are to be avoided.

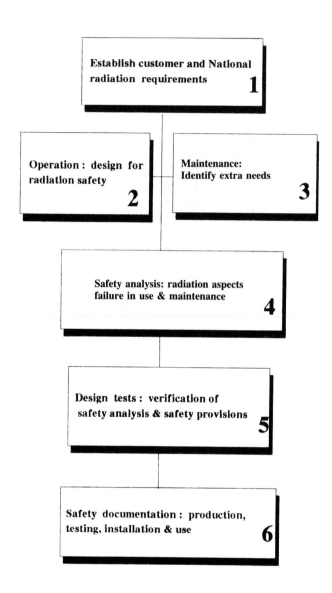

Figure 10.2   *Flow diagram – product design considerations*

Figure 10.2 is a sample flow diagram illustrating some key elements of those radiation safety considerations necessary for the design of products producing sufficient RF energy to require a safety assessment. Most of the general points will also apply to any consequential X-ray radiation produced. The diagram can be applied both to the provision of a design for a particular customer and to the design of products aimed at a general market.

The obvious difference when interpreting the notes which follow is that in some cases there is a specific customer to talk to, whereas with speculative design for general markets, the marketing and sales functions have to answer the questions implicit in headings 1 and 2 of the flowchart. The latter case may involve meeting the tightest radiation safety requirements existing in the selected market area.

Since the intention is that the steps identified in the flowchart should be integrated into the design plan for equipment which is being designed, items such as the radiation safety analysis form just one element of the total safety analysis for the product.

The flowchart also applies to RF sources other than radio transmitters, for example RF process machines and RF sources for medical use, the degree of applicability depending on the nature of the product.

Customer and national radiation protection requirements

Some customers, particularly large ones, have corporate or local requirements and standards to be met which may be reflected in a specification or local management document.

These may include the standards used for RF and ionising radiations and, importantly, a conservative organisation may have tighter requirements than the national guides and regulations.

It is clearly essential that there is no misunderstanding about this, since remedial action after delivery can be a very expensive proposition and is also a poor advertisement for the supplier. Some large customers have specific test methods to be applied to the radiation acceptance tests on the delivered equipment which may involve additional test equipment, compared with that normally used.

As far as national requirements are concerned there are variations between countries in the approach to both RF and X-ray safety. Sometimes regulations are accompanied by codes of practice which may, as in the UK, have a legal status. In this code of practice there is one section for the product producer and another for the product user.

If a transmitter produces X-rays it generally falls into the category of a 'radiation generator' subject to similar legal provisions to those applying to X-ray machines and other similar sources. It is clearly important to take into account all the legal requirements in those countries constituting the proposed market.

Some countries even have requirements for the calibration of measuring equipment to be performed either in that country or in one with which they have reciprocal agreements. Failing to become aware of such requirements can result in considerable expense and lost time. Most countries do require the formal calibration of ionising radiation measuring instruments.

Certain countries may also require formal 'type approval' of some types of transmitter, covering all aspects of the product design and performance, especially ship transmitters and the like.

The European Economic Community (EEC) already has a Directive on Ionising Radiation (the Euratom Directive) and is currently developing a Directive on Physical Agents, which includes requirements covering RF and other types of radiation. There is also a further Directive on Machines which takes in RF processing machines.

The broad basis of the provisions in the twelve countries will therefore be the same. It should be noted that EEC Directives are implemented in the national laws of those countries. Consequently, some of the detail of the implementation may be expressed in different ways.

Using products

Personnel who operate transmitters or other RF sources need to be protected against excessive RF radiation exposure and, where applicable, against X-ray radiation. Here 'excessive' means in excess of the relevant specification requirement, standard or legal provision. Protection should be related to the supplier's safety instructions provided with the product.

For example, if the operator has to carry out an adjustment or other manipulation with a radiation shield removed or a door opened then some provision is needed to compensate for the loss of shielding resulting from opening the door or other shield (see below).

The aspects most relevant to the operator include:

1 Leakage reduction from the transmitter cabinet, feeders, antenna matching units and all the associated items which constitute a system or installation. Both X-ray and RF should be covered where applicable.
2 Adequate information on the radiation safety aspects of the equipment, including any prohibitions applying to the opening of inspection and other panels.
3 The ability to identify feeders and other associated equipment by means of unique markings, so as to avoid any hazard resulting from accessing the wrong items.

Some national product safety requirements expect designers to take into account possible minor misuse of products. This includes measures to reduce the possibility of accidental misuse due to inadequate instructions or other causes.

Product maintenance

1 Often during maintenance, the range of potential hazard is increased due to the diagnostic techniques involved in determining what is wrong with the equipment. It is important that maintenance instructions provided by the supplier should provide for adequate safety in respect of radiation.

A typical case may be one where the safety interlocks on a door are required to be overridden in order to perform a maintenance adjustment.

If opening the door to perform the adjustment reduces the protection to the maintenance technician, then a temporary servicing screen or other device should be provided so that the protection is restored.

2 Maintenance engineers are inclined to remove parts in order to get a better view of what is wrong. They then switch on again and observe the results.

Where X-ray radiation is concerned, it is essential that adequate signs and maintenance warnings exist to prevent the maintainer taking out the X-ray shielding including any fortuitous or intended shielding provided by other metal objects and structures which are nearby. Similar RF limitations should be made clear. Marking warnings on shields is more effective than a pure reliance on maintenance handbooks since people tend not to resort to the latter until they are in difficulties with fault finding.

3 Since practical installations of transmitters may involve long feeder runs, antenna exchanges, dummy loads and similar ancillary items which often need to be moved or operated in some way, as well as to be maintained, it is important to ensure that inspection plates and other access provisions on these items bear adequate warnings to prevent inadvertent access to RF energy. In very high power systems it may also be necessary to consider electrical interlocking as well.

The risk tends to become greater when working on equipment which is not close to its parent transmitter since when there are, for example, a number of feeders running together, it is easy to pick the wrong one unless they are clearly marked, due to difficulty in following the runs visually.

4 The supplier provides maintenance instructions for the user and it is important that the range of known maintenance requirements are considered and any design actions relating to them are documented. The writer of maintenance instructions, or better still a different person associated with the design activity, should always try carrying out the draft maintenance instructions, following them literally. This could result in a significant improvement in maintenance instructions!

Radiation safety analysis

Having looked at the design requirements, the purpose of the safety analysis is to examine the adequacy of the proposed design against those situations which might be expected to occur in use and in the maintenance activity, including any radiation hazards which can result from an equipment failure.

Until the equipment reaches the testing stage it will not be possible to be certain of efficacy of such things as the proposed shielding material thickness but the protection intended to be provided can be recorded against the circumstances envisaged e.g. normal operation, maintenance adjustment, etc. It can be useful to list particular items of relevance such as doors, removable plates, windows and to identify the corresponding safety provision e.g. interlocked, marked with a warning, lead glass window used.

Where the design is a project for a specific customer there may be more definitive safety aspects to be considered such as the contribution of the customer to the final system. This might include wiring, buildings and utilities. The effectiveness of RF earthing is often in question. There are situations where an inadequate approach to RF earthing may provide an RF burn hazard at lower frequencies.

Design tests

It is very important to ensure that design tests include thorough testing of all safety features and explore all the aspects outlined in the previous paragraphs. They should be carried out in the worst cases likely such as maximum power output, likely worst frequencies and modulation depths. For pulse transmissions the worst case pulse characteristics should be explored.

Particular care should be taken to test whether safety circuits are likely to be interfered with in an adverse way. It is not unknown for RF interference to operate fire detectors of the electronic variety and these are sometimes built into transmitters or transmitter installations.

Similarly, digital or other circuitry used to control interlocks and other safety systems should also be checked for spurious operation over a number of frequencies if the transmitter has a wide frequency range. Sometimes only one frequency or one part of the frequency band may cause interference with such circuits.

The effectiveness of any windows fitted to the transmitter should also be specifically checked. Lead glass windows need to be checked for X-ray protection and ordinary glass windows with any associated wire mesh for RF reduction should be checked for RF leakage. General leakage testing should include checks on contact strips around doors and for leakage at door handle catches which often prove to be a leakage point.

Not all safety equipment in high power transmitters is aimed at personnel protection as some is required for the protection of the equipment. Such items include overload detectors, thermal monitors and flash-arc detectors to protect against flash-arcs in high power electronic tubes. It is equally important that these items are free from any adverse effects caused by stray

RF radiation. The consequences may be economic e.g. changing an expensive high power tube when the fault is in the flash detector and not the tube.

Design safety documentation and records

It is important, both for compliance with national health and safety requirements and for proper product quality management, to keep a good record of design safety features, any tests carried out at the original design stages and after any modification which could possibly affect radiation safety.

Because of the great importance of the product modification activity and the need for a continuation of the quality management of the product, it is usual to include some form of declaration on modification documentation so that someone suitably qualified to do so certifies whether radiation safety may be affected or not. If in doubt, a positive assumption should be made and tests carried out.

Modifications often pose considerable risks in so far as the solution to, say, a functional problem which is, understandably, the preoccupation of the design engineer, may bring with it unexpected side effects. Records of changes in safety features and the safety test results should be retained on a systematic basis so that the product safety record at any point in time provides a cumulative picture of the care taken to ensure human safety.

Technical aspects of design

Shielding

General

The common objective of X-ray and RF shielding is to confine the radiation to the cabinet, case or box surrounding an RF source so that the only radiation appearing externally is the wanted output to an antenna, applicator or work piece. The materials used to do this differ in their response to the two types of radiation. Basically shielding materials may absorb and reflect energy. In the case of RF energy, most is reflected back into the equipment.

RF absorber materials may also be used at higher frequencies, the object being to minimise reflection by absorbing energy (see later).

Where modern plastic-based RF absorber material is used it is important to ensure that it cannot be thermally overloaded. Excessive temperature may lead to fire and to the production of very toxic vapours from some materials.

The choice of materials for shielding needs to be both appropriate for the type of radiation and compatible with other metals used from the point of view of avoiding galvanic corrosion. For X-ray shielding, discussed in more detail later, the aim is to absorb the X-ray energy and material is chosen for its attenuation properties.

The objective common to both types of radiation is for a transmitter design to attempt to confine or attenuate RF and X-ray energy as close to the source as possible. There are, of course, technical problems with shielding X-rays at the source e.g. final power amplifiers, due to the capacitance added in the circuit if metal shields are brought close to these tubes. Nevertheless the basic aim should be borne in mind since the cost of dealing with leakages occurring at a number of places in a cabinet can be considerable, especially in respect of X-rays.

Cabinet design

1 The design of cabinets is obviously a key factor. Inset (flush) cabinet doors will tend to be better than doors which project outside the cabinet, i.e., externally hung doors. This is because the cabinet around inset doors provides extra shielding at the door edges.

2 'Spare' holes in panels or gaps between panels, especially in equipment where various panel or sub-panel optional items may be purchased, should be avoided by the use of blanking plates, etc.

3 Small inspection flaps and covers should be firmly fixed so that tools are needed to open them and radiation warning markings put on them.

4 Glass windows (usually provided to observe electronic tubes, RF arcs, etc.) should be shielded with a wire mesh to reduce RF leakage.

5 Lead glass or equivalent should be used for such windows when X-ray leakage is involved. Wire mesh may still be required for RF leakage as lead glass does not provide an effective RF shield.

6 Where lead glass windows are used, the equipment drawings and the user handbook should carry suitable information. In the case of drawings, the window should be so closely defined as a purchased item that no inadvertent substitution with ordinary glass can occur. The need to inform the user is from similar considerations.

7 Where the transmitter output is via waveguide and there is a significant amount of X-ray generated, a lead gasket is often used at the tube flange.

8 A careful choice of earthing strip for cabinet door sealing is needed to ensure proper contacting without arcing. In particular, for microwave frequencies where the wavelength may be of the same order as the usual fixing holes for earth strip, fixing spacing distance and contact finger design should be such as to avoid possible resonant dimensions.

9 In high power HF broadcast transmitters, the earth contact strip used round doors can give rise to arcing and, unless substantial enough, the heat can cause some oxidisation and surface discoloration often resulting in increased contact resistance and subsequent greater heating.

Thin film plastic-backed contact strip materials, which may be suitable for low level signals, are likely to be unsuitable where there is significant energy since arcs may melt the backing and glossy materials may be heated by the leakage field.

10 Occasionally a transmitter might be located above ground level, for example on the first floor. Now most high power transmitters do not have a 'bottom' in shielding terms since they are traditionally located on the ground floor with a substantial concrete base. If such a transmitter has significant X-ray energy, there is always the possibility of an X-ray beam down to the ground floor. There may also be some RF radiation reaching the ground floor. Consequently, it is desirable to identify such aspects at the earliest possible stage.

11 Air ventilators and air filters can be a weak spot for leakage, depending on the nature of their structure and the materials used. Consideration of this aspect before manufacturing or purchasing such items will avoid unnecessary expense. In the case of RF leakage, a suitable wire mesh across the filter may again solve such problems. Proprietary honeycomb materials are also available.

Safety interlocks and safety systems

1 Equipment with high voltages present will have interlock systems to prevent electrocution. Where these interlocks are intended to act also as a protection against radiation, this needs to be made clear in handbooks and the like. In maintenance, the practice of overriding interlocks is prevalent and in such cases, as noted earlier, servicing shields may be required to restore protection. In many countries the failure to provide proper shielding and interlocks, where necessary for protection against ionising radiation, may constitute an offence.

In cases where the structure of a transmitter is such that people can physically enter it via doors which are interlocked, it is necessary to consider provisions against failure of the safety systems such that a person may become exposed if the equipment is then switched on whilst inside.

If X-ray radiation is present there is usually a legal requirement for the provision of a manual alarm to be operated by the trapped person. Commonsense suggests that this should always be provided since there may, in any case, be electrical and RF hazards.

2 As mentioned earlier, RF fields can give rise to the spurious operation of some types of safety equipment and circuitry within the transmitter or the associated system.

Often those who design such circuits may have a role which does not give them a great deal of familiarity with RF radiation. Thus the transmitter designer should communicate such requirements and also ensure that the layout of transmitters and associated systems take into account the need to ensure the integrity of all elements of the safety system. Wherever possible, safety circuits should fail safe.

3 Interlocks in rotating antenna systems should be so arranged that rotation can be positively inhibited, preferably by a good quality key switch. Provision for the stationary operation of moving antennas is almost always required for survey purposes and should be provided for in the design, as 'standard'. There should also be an audible warning signal sounded before movement starts when rotation is switched on again.

4 Similarly, where applicable, provision should be made on rotating beam systems for a sector blanking facility. Where this is done by software, some thought needs to be given to the protection of authorised settings e.g. by 'password' access or some equivalent. Where the design allows, a removable key switch may be even better.

For all mechanically moving antenna systems, arrangements should be made in the design so that if movement fails, the RF radiation is switched off automatically. This prevents a failed system continuously irradiating people and is, by definition, a prerequisite for the assessment of moving beams by time-averaging, if otherwise exposure could exceed permitted limits in the 'failed rotation' situation.

Safety indicators

It has been found advantageous in many situations to associate a flashing lamp with the operation of high power transmitter systems by the provision of a circuit which is only energised when the antenna is fed with RF energy. This serves a number of purposes including distinguishing moving antennas being operated only for mechanical tests or running in bearings, from those which may have RF hazards. It also helps with surveys in indicating that the transmitter is still operating. The lamp needs to be located close to the antenna.

There may be a similar case for this on RF process machines though, as with X-ray machine installations, a steady red light may be more suitable as a flash may cause irritation to people because of their relative closeness.

Components

Where there is considerable X-ray radiation present in a transmitter, it should be noted that there is at least a suspicion that this can cause the failure of important components such as vacuum capacitors and some plastic dielectric capacitors. Whilst this has not been proven, prevention is better than cure! Prevention may involve careful siting and possibly the shielding of such items. Generally speaking, a metal shield will be much cheaper than the component being protected.

The suppliers of electronic tubes will usually provide some information on the magnitude of the X-ray doserates likely under particular conditions. See Chapter 7 for some examples.

Warning signs

X-rays

The X-ray sign is international and consists of the trefoil in black in a triangle on a yellow ground. The triangle is defined by an ISO standard as a 'caution' symbol. The symbol is given in BS 3510:1986[58] and in ISO 361.

RF radiation

In the UK a symbol for non-ionising radiation has been established for many years and is a triangular symbol consisting of a black radio tower emitting waves, on a yellow ground.

It is defined in BS 5378 Part 3[59]. Part 1 of this standard deals with the general provisions for safety labelling and is invoked in UK law. There does not seem to be an international symbol although this would not seem to be a controversial matter.

Other warning signs which do not expressly relate to radiation but reduce the risks when measuring radiation include the identification of components which get very hot, signs indicating the location of hazardous substances and the general identification of individual antenna feeders and ancillary devices.

Vehicle installations

1 Mobile equipment of the vehicle mounted type can need particular care. HF equipment using whip or extendible rod antennas can, if of sufficient power, offer a number of hazards. These include the possibility of contact burns from bare metal on the vehicle structure and from the antenna and feeder, as well as personal exposure to the field.

   Sometimes the dimensions of parts of the structure are such that they are driven parasitically from the antennas on the vehicle and there can be a potential for contact burns on such parts as the corner vertical members of small van bodies. Side mounted antennas fixed vertically with the feed point at perhaps waist height for a standing adult provide the possibility of burns and of exposure in excess of permitted limits.

2 Where such vehicles are used in the public domain consideration needs to be given to the avoidance of such exposures and the reduction of avoidable burns, where possible, by the use of insulating shields and the like. Antenna feeder connections are obvious candidates for physical shielding and transparent materials can be useful there. The reduction of exposures

to the public should generally be achieved by suitable siting when operating.

3 The exposure for those operating the equipment in the vehicle can be much reduced by careful arrangement of the equipment and the antenna feeder runs. Within a vehicle it is usually beneficial to run antenna feeders away from the equipment operator e.g. behind the equipment rather than in front of it. Feeders run in the roof of a vehicle should not be routed over the heads of the operator or other seated people, if this can be avoided as it usually provides unnecessary psychological problems in respect of small amounts of field, which can easily be avoided by suitable routing.

Sometimes some shielding of cables might be needed. Also, an excess of earthing points can increase the field inside the vehicle so care is needed in the way earthing is implemented.

4 Where roof ladders are fitted to vehicle installations, suitable warnings may be needed when the antenna systems are roof mounted to prevent anyone climbing the ladder unaware of any roof hazards.

5 Remote control lines, telephone and other wires for local functions can become subject to induced RF voltages and filters may be necessary to avoid burns to personnel and also possible damage to the remote control equipment. The latter is perhaps less likely on long lines due to line capacitances.

Ship systems

It is difficult to generalise about ships since they vary enormously in size, nature and function. A few frequently recurring points are worth noting.

MF and HF installations

At the lower frequencies, the electric field associated with the feeders, which are often made of copper tube, can be very high. Some improvement can be made if the feeders are kept away from the operator. In some cases shielding of the feeders may be needed to meet the requirements of safety standards.

Care is needed with the location of the antenna system, though there is often not much choice. The main hazard apart from the antenna itself, is the creation of hot spots around the deck where shocks and burns can be incurred due to resonant metal structures being energised parasitically.

Other systems

Not a great deal can be done about the design of most equipments for this environment as the predominant hazards on board are:

1 Exposure from a variety of antenna systems when moving about on deck and when climbing.
2 Risks to flammable vapours.
3 Shock and burns, as already noted.

Apart from the siting of antennas, none of these points can be influenced by equipment design.

X-ray shielding data

General

The most general residual problem with equipment which produces unwanted X-ray radiation is not usually a steady excess of X-ray leakage, although this can happen if shielding design has been neglected, but rather, a number of leakages from holes, and gaps, door catch fittings and similar places.

There is, worldwide, a considerable amount of published data on the X-ray attenuation characteristics of both metals and other materials such as brick, concrete and similar building materials, largely stemming from the X-ray machine field. In the UK there is a British Standard covering X-ray shielding BS 4094:1971 part 2[53].

The attenuation of a given material follows an exponential law:

$$R = R_o\, e^{-\mu x}$$

Where R is the intensity transmitted through thickness 'x' of the material;
R_o is the initial intensity of the beam;
x is the thickness of material;
e is the mathematical constant e;
μ is the total linear attenuation coefficient of the material for the energy of the particular radiation involved.

The quantity 'μ' has to be found from a table of values.

The attenuation in decibels will be $10 \log_{10} R/R_0$.

Published tables and graphs usually plot attenuation versus material thickness for different energies or peak voltages. Some data from the X-ray machine field may have graphs for constant voltage and pulsed voltage tube supplies.

Two particular concepts are of use in shielding work. These are the half value layer and the tenth value layer of a metal for a given energy and are defined in terms of the thickness of the metal. They are defined as follows:

'Half' and 'tenth' thickness layers of shielding materials

Half-value layer thickness (of a material)

The half-value thickness of a given shielding material is that thickness which, when interposed between the source of the X-ray beam and the measuring instrument, results in the doserate being reduced to half the previous value, when measured at the same place.

Tenth-value layer thickness (of a material)

The tenth-value thickness of a given shielding material is that thickness which, when interposed between the source of the X-ray beam and the measuring instrument, results in the doserate being reduced to one tenth of the previous value, when measured at the same place.

Measuring the effective energy of X-rays from a source

In order to measure the effective energy of the X-rays from a source to provide information with which to select the required thickness of shielding material, the method used involves half value layers of appropriate metals.

The basic concept may be unfamiliar to people who have not been involved with X-rays but is straightforward. If we have a beam of X-rays being measured by an X-ray doserate meter and set the position of the meter to give a convenient reference level, we can now insert plates of suitable metals of differing thicknesses in the beam one at a time until the meter reading falls to half the initial reading.

The plates should be large enough to ensure that all the beam goes through the plate. The plate which causes the meter to fall to half the previous value is known as the half value layer or thickness for that energy. If reference is then made to a table of thicknesses of the metal used against the energy for which that thickness is the half layer value, then that figure is the effective energy of the beam.

The method is a well-established method of measuring effective energy and half value layer thicknesses can be obtained from X-ray reference data tables. The method is also incorporated in IEC standard IEC562[57]. In that standard there is a table of metal thicknesses and effective energies which can be made up as standard test plates (see Table 10.1). Figure 10.3 is a plot of the Table 10.1 thicknesses versus energy up to 100 keV.

The test plates can be made of the metals specified and the thicknesses given in the table, each plate being marked with its energy value. For example, a plate of aluminium 1.1 mm thick can be marked '23 keV effective energy'. A set of plates appropriate to local needs can thus be made as standard test equipment.

Table 10.1   *IEC Standard IEC562 table of metal half value layer thickness versus energy (Copyright the IEC Geneva)*

Effective energy (keV)	Material	Half-value layer (mm)
7	Aluminium	0.1
21	Aluminium	0.9
23	Aluminium	1.1
35	Aluminium	3.3
50	Aluminium	6.6
50	Copper	0.3
80	Copper	1
100	Copper	1.4
130	Copper	2.7
160	Copper	3.9
200	Copper	5

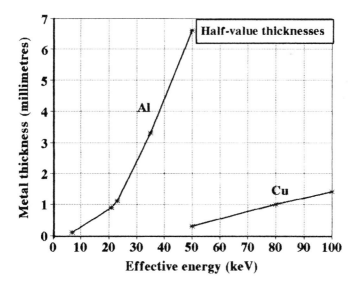

Figure 10.3   *Half value layer thicknesses plotted from Table 10.1 (Courtesy of the IEC)*

Unfortunately, because a particular series of values of effective energy was chosen when the IEC562 standard was created and the metal thicknesses then defined to meet the energy values listed, some of the resulting metal thicknesses are not commercially available. Interpolation can be used to suit available metal thicknesses. Apart from this problem, the method is easy to use.

Values for the tenth value layer thicknesses have to be found in other reference data, as does the half value layer for energies not covered by the IEC standard. Figure 10.4 gives an example of the half and one tenth values

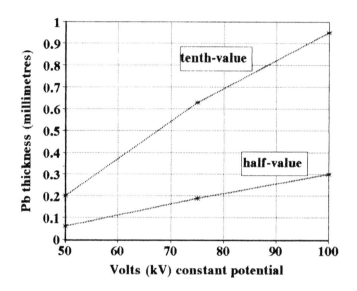

Figure 10.4 *Half and one-tenth value layer thicknesses for lead shielding*
(Courtesy BSI, London)

for lead, from X-ray machine data for constant potentials in British standard BS4094 part 2[53].

Figure 10.5 gives the tenth value for aluminium, also from the British standard. These curves are best considered as illustrative and more specific data used where possible, e.g. from other individual graphs in the standard. Standards like BS 4094 usually give plots of transmission (i.e., attenuation ratio) versus material thickness for different values of energy or tube voltage. The flatter portion of such curves should be used.

When undertaking X-ray measurements it is often useful to be able to estimate the thickness of metals needed to overcome a leakage problem by experimentation.

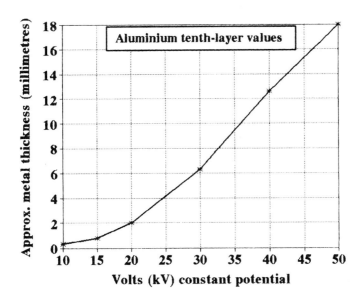

Figure 10.5 *Tenth value layer thicknesses for aluminium shielding (Courtesy of BSI, London)*

On an experimental basis it is possible by using a number of plates to make very rough estimates. Suppose that, using the data from the earlier example, several plates 1.1 mm thick are available and the X-ray effective energy has been established as 23 keV. Each plate is a half-value thickness for 23 keV so that two plates used together will give an attenuation of 0.5 × 0.5 = 0.25 (one quarter), and so on. If some plates are available at the tenth-value thickness for 23 keV then we can go further and have attenuations such as 0.5 × 0.1 = 0.05 (one twentieth).

The reason that this was described as an approximate method is that as the metal sheets increasingly attenuate the low energy X-rays, the attenuation of the high energy part of the energy spectrum is appreciably less and the effective energy of the residual leakage becomes higher as illustrated in Chapter 7. Hence the plates no longer correspond to the half and tenth values for this higher energy.

Consequently, it is usual to estimate material thicknesses required for either initial shielding or for remedial shielding very conservatively, perhaps by assuming a higher effective energy than that measured with a single half value layer. It is usually cheaper to buy a thicker material than is strictly needed rather than to order a thinner material, wait for delivery and find that it is not quite good enough.

Lead glass

Lead glass is commonly used where a window is required for observation of an electronic tube or to look for arcing, and there is X-ray radiation present. It is extremely effective providing that the glass is properly mounted so that there are no gaps around the periphery. Figure 10.6 shows graphically a

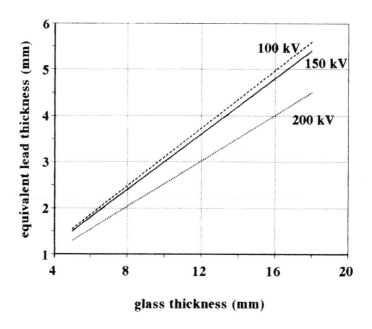

Figure 10.6 *Curves for lead-glass shielding material (Courtesy of Pilkington Special Glass, Wales, UK)*

manufacturer's published data[54] for a proprietary lead glass. It is used here for illustration.

Similar glass is available in other countries. When contemplating using such data, the manufacturer's current data should always be obtained as there can be changes from time to time, and different manufacturers may have different data. In this example, the data on attenuation is given in terms of peak voltage versus 'lead equivalent'. It is necessary to consult a lead table or chart to establish the effective attenuation.

In choosing glass it is necessary to remember that standard lead glass is made in specific sizes and these can be marked on the 'glass thickness' axis of Figure 10.6.

RF shielding materials

For many large transmitters the materials used will often be those of the cabinet structure e.g. steel, and aluminium. Other common materials include copper and brass. Comprehensive design data can be obtained from specialists in this field. The proprietary materials mentioned here are from a UK supplier's catalogue[55] and are available world-wide. Similar materials are also available from other suppliers. Reference 21 gives data on the shielding properties of materials used for RF applications.

Contact strip for cabinet doors on large equipment needs to be robust. Beryllium copper is usually used. Both standard and custom-made sizes and patterns are widely available.

Wire mesh materials are readily available for shielding windows. These materials are all well established and characterised for electrical purposes. The performance depends on the the frequency, the nature of the wire mesh and the aperture size. Reference data is available on the attenuation of such materials both from reference books and from the many manufacturers of EMC shielding materials. Proprietary ready-made window fabrications using wire mesh sandwiched between glass or acrylic plates are also available, together with performance data.

With the extensive use of plastics in electronics, there is a large range of possible approaches to shielding. Shielding by vacuum deposition or sputtering is in common use on plastics, ceramics and glass substrates. Coatings include pure aluminium, silver, nickel and copper. Multilayer coating, particularly with stainless steel and copper are said to give very good shielding.

It is often important to have shielding which is transparent, especially for use with RF process machines where the operator needs to watch the process. Again, glass, polycarbonate and acrylic materials are available with shielding properties provided by the deposition of a fine layer of conductive material such as indium tin oxide.

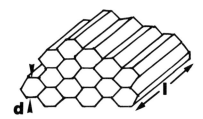

Figure 10.7 *Honeycomb vent utilising waveguides below cut-off*

This is said[56] to have no noticeable effect on the transparency of the material in a very thin layer for electrostatic purposes but when used in thicker layers for electromagnetic shielding, transparency is reduced by between 5 to 35% for sheet resistivities in the range 8 to 20 ohms per square and the material has a yellow hue.

Other ready made and made-to-order shielding materials include gaskets, woven meshes, air vents, etc. Of particular interest are the honeycomb air vents mentioned earlier. These are generally multiple hexagonal cells in aluminium which provide RF shielding by operating as a waveguide below cut-off. Figure 10.7 gives an impression of the structure of honeycomb structures. The length of the cell needs to be four or more times the diameter of the cells ($l/d \geqslant 4$). The open cells offer a low resistance to air flow. Dust filters can be used with these devices.

11
RF radiation safety management

Safety management objectives

Over the last twenty years or so there has been a considerable increase in the general awareness of the hazards around us, both at work and in the social and recreational environment. Traditionally, safety management has long had an important role in the more tangible aspects of life such as the machine safety and structural safety fields. Ionising radiation legislation and management has also grown to a highly regulated state with basic guidance emanating from an international body so that there is some degree of universality particularly where nations interact with each other, as in the transportation of radioactive substances.

Whilst the safety hazards connected with RF radiation have been subjected to some degree of control in most countries, it is a subject which is not always understood. Unlike the position with ionising radiation, it has not yet found a generally accepted international basis. This is mainly due to the lack of consensus in views on the subject.

The desirable objectives of safety management in the RF radiation field are, in the generality, much the same as for any other subject. They include:

1 Identification of potential hazards to people in the working environment and to nearby people in the public domain. Where high voltages are used, any co-existing X-ray radiation must not be overlooked.
2 Providing methods to reduce such hazards at least to the limits and levels provided for by the relevant laws, codes of practice and other national provisions.
3 The provision of, or the use of, competent measurement services to determine and record the measured values of those quantities used for controlling safety.
4 The education of technical personnel and others in the safe use of RF radiation both in relation to intended and unintended radiation.

5 The development of the radiation safety features of product designs to reduce the potential hazard to users.
6 Investigation of alleged over-exposures.
7 Maintenance of the records of the measurements made, safety provisions instituted, complaints received and the action taken to deal with them.
8 Control of access to RF fields in general including the policy to be operated in respect of personnel and visitors who have been fitted with implanted devices such as heart pacemakers.

The nature of these responsibilities and the knowledge and experience required will depend on the type of organisation concerned and these can be categorised, if somewhat loosely, as follows:

Companies designing, manufacturing and installing sources of RF radiation

Here there is a responsibility for the design and delivery of safe products and the knowledge required to tackle the safety aspects will often be considerable. This will include the writing of safety instructions for the user and producing suitable maintenance instruction manuals.

Companies who manufacture other people's designs

In most countries, the legal responsibility to deliver safe products will still apply in this case and the manufacturer will need to ensure that there is adequate expertise available to meet his legal obligations.

Similarly, in many countries, importers of equipment will have full liability for product safety in use and when being maintained and will need to have a good knowledge of the way the product designer and producer have investigated the safety aspects of the product.

Equipment users

This category will include personnel on transmitting stations, those using mobile equipment by land, sea and air and those in charge of RF machines for medical, production and other applications. The safety provisions may have to take in an aggregate of equipment constituting a system and if none of the suppliers has any overall system responsibility, then the purchaser has, effectively, taken on the system safety responsibility and must have sufficient knowledge and experience to handle it.

Equipment users may include anyone who is licensed to radiate RF including non-technical users of mobile and fixed equipment. Some organisations e.g. public services, will have technical management staff to be responsible for safety and to guide non-technical users. Others may be dependent on the adequacy of the supplier's instructions.

The identification of potential hazards

The earlier chapters have dealt with the technical aspects of the various hazards attributed to RF radiation and it may be useful to summarise these so that the relevant aspects can be recognised on a checklist basis. Some of these items may only rarely be experienced or not experienced at all.

1 Human exposure to RF fields

- Exposure of all or part of the human body to RF fields (referred to as whole-body or partial-body exposure).
- Exposure of the human body to peak pulse power densities.
- Limb current limitations.
- RF burns and shocks.
- Aural effects (hearing the pulse repetition frequency or noises and crackling associated with pulsed radiation). This is not regarded by itself as a hazard, but may be so regarded by those who experience it!
- Effects on implanted devices such as heart pacemakers, insulin pumps, other active devices and passive conductive materials.

2 Human hazards indirectly associated with RF radiation

- X-ray irradiation from high voltage electronic tubes.

3 Effects on materials and substances which may or may not involve people.

- The ignition of flammable vapours and electro-explosive devices by RF fields.
- Damage to static sensitive devices (integrated circuits and the like) being assembled in the proximity of a strong RF field.
- Electromagnetic compatibility (EMC) problems where interference with other objects or their electronic systems ranging from aircraft to motor car electronic control systems may cause problems. The consequences may be minor or very serious.

4 The effect of static (DC) magnetic fields on the human body

- It may happen that such fields are present in some equipments which need to be checked for RF radiation, particularly in the medical field.

  Where there are standards in force which specify separate and different safety limits for occupational purposes and for the public, it may be necessary to consider such of the items listed above as are relevant, under each heading, so as to ensure that both situations are satisfied.

Safety management methods

Resources

Safety management requires adequate resources to undertake the required functions, a fact not always appreciated. This is equally the case where RF radiation is involved and where some specialist knowledge and experience is inevitably needed.

Since much of the basis of RF safety management is related to appraisal and measurement, then adequate equipment resources are necessary. Whether this is achieved locally or by the hiring of a measurement service is not, in principle, of great importance providing that the requisite knowledge is available. The relative practicability of the two will generally be determined on the basis of the required response time for the provision of services.

Equally as important as the measurement resource, is the management provision which ensures that all the relevant equipment and systems do get surveyed when this is necessary. This is just as important in large organisations which design and manufacture RF radiation sources as it is in user organisations.

Unless there is a good system to ensure that all equipment, or a sample of production if this approach is used, is surveyed, it is possible and has been known for a new design to get into production and almost out of the factory without this having been done.

Similarly it is not unknown for an RF process machine to get into use without being checked out for leakage by the purchaser. Sometimes the first intimation of the arrival of such a device in the past has been a complaint about something interfering with the local computer!

Although in the production of modern RF process machines one can reasonably expect that thought has been given to RF leakage, there is clearly a need for the purchaser to meet the legal requirements for ensuring the safety of his personnel.

Procedures

There are many good reasons why there should be adequate documented procedures for the management of RF radiation safety. The problem with procedures arises when managers believe that what is written will be always be done! Almost every accident investigation highlights breaches of procedures.

Procedures work best where the activity is a boring one and where the reader is only too glad to follow it. For example, procedures to complete forms, for report formats and similar mechanistic activities usually work reasonably well. Detailed procedures also work well in areas of activity where strict disciplines are the normal practice and the procedure documents are always available, as in calibration laboratories, and in groups of people specialising in measurements such as RF surveys.

In such groups there is generally a heightened awareness of the importance of the quality conformance aspects coupled with an understanding of the problems which may stem from a departure from such procedures. There is also a sense of direct personal responsibility for the consequences.

Complex technical procedures for a wide general use are difficult to remember, and the written copy is often not to be found when needed. This leads to improvisation.

It is unfortunate that the most helpful people often prove to be the greatest safety risks, since they will try anything whether or not they have the requisite knowledge. The helpful person who hastens to pick up a dropped radioactive source with his bare hands is a good example!

In the author's view, global procedures which relate to specific technical safety matters such as prohibiting access and aimed at wide general use should be regarded as high risk and wherever possible more positive methods of prevention should be used. For example, a 'fold up and lock' ladder which prevents the climbing of towers is far better than a procedure forbidding it.

Of course many topics may need to be covered globally for legal reasons and because they fulfil the quality management system need for documentation. The main point made here is not that one should not bother with such procedures – they are very important, but rather that wherever possible technical methods should be used to make it rather difficult to ignore or disobey critical aspects of the instructions.

An equally important aspect is the communication of procedures to all concerned. Positive methods such as safety briefings, the inclusion of safety training modules in company training courses and the like are necessary to overcome the reluctance of people to read manuals. As with many other subjects, safety training should start at the top of the management tree, not work its way up from the bottom!

Those written procedures which are certainly necessary include those covering:

- Safety policy.
- Training.
- Measuring equipment maintenance and calibration.
- Reporting and investigating alleged overexposure.
- Records and record keeping.

Procedures are also likely to be needed for such things as:

- When surveys are done, who can initiate surveys, and the method of reporting on them.

- Technical methods of dealing with excessive radiation.
- Definitions of prohibited areas, 'permit to work' systems.
- Anti-climbing provisions, authorisation of climbing, etc.
- Product design safety reviews (where design is undertaken).
- Control of works on site, including sub-contractors.

Handling survey reports

Surveys should be initiated from time to time when changes in equipment or operating conditions arise or when a periodic check is thought desirable. Survey reports provide measurement data relating to particular situations. They will sometimes highlight problems and provide some recommendations for remedial action.

The main safety management action relating to survey reports is to ensure that the necessary actions are taken to clear any deficiencies in the present situation. Since there will often be options for remedial action with different time and cost implications, technical discussions may be needed to determine the specific action.

There is also likely to be a need for temporary action if excessive exposures have been established. This may involve quickly applied measures such as the use of a temporary barrier around the area concerned and an access prohibition notice. The essential management action is to pursue and progress short and long term actions until completed, recording the completion in some appropriate way.

The actions involved may be physical (barriers and signs) or procedural (e.g. establishing a negative elevation limit for a microwave beam).

Some surveys will be carried out with the specific objective of setting safety limits prior to the deployment of personnel. In this case the surveyor has only to define the safety limits. Problems only arise if these boundaries are found to be incompatible with the intended personnel deployment.

Survey reports and the records of any remedial action taken are vital safety records and are best retained indefinitely. In organisations where a lot of surveys are undertaken it has been found valuable to index them against equipment types so that results at different locations with the same type of transmitter system or RF machine can be cross-checked.

Safety audits

Since the use of safety procedures are so liable to fail with time or with changes of personnel or merely because management does not seem to be very interested, the carrying out of safety audits from time to time is very necessary. Whilst most readers will be familiar with the concept of an audit since it is a common quality management term, it is perhaps worth a brief explanation. An audit is simply an objective look at a given situation in relation to the rules, instructions and practices which are intended to control that situation.

The outcome is a report which identifies any situations where there were breaches of procedures, instructions and practices. It may also identify situations for which no adequate control instructions appear to exist but are thought necessary by the auditor. It may also identify ambiguities in existing instructions and procedures and note suggestions from the personnel concerned. The report gives the local manager the chance to debate the issues and agree remedial action.

Safety audits are usually carried out in such a way that every area of work is covered over a given period, the timing and frequency being determined in the light of the nature of the organisation. The most important feature is the requirement for a definitive response from the manager of the area concerned in a given period of time.

Safety record keeping

It is very important to keep full records of all surveys and the action taken to deal with the results. Similarly, records of permits-to-work and authorisations to climb towers and structures should be retained. Records of minute and negligible readings found are just as important as records of high readings as they establish that care was taken and that no hazard was found.

Most claims against organisations for alleged injury due to RF and other radiation are made many years later. The long term retention of all survey records is therefore crucial. A lack of records is hardly likely to help the case!

Records of safety audits and the remedial action taken are equally important since they can provide evidence of the care taken in safety management. It has to be said that when not taken very seriously they can also prove to be a record of inadequate safety management!

Records of people who have been involved in alleged over-exposures of any kind of radiation together with the results of the investigation should be kept indefinitely. It is wise to do the same with external complaints from the public and the resulting investigation results.

Where people are provided with thermo-luminescent dosemeters (or the film badge equivalent) for X-ray dose recording, as a prudent safety measure, i.e., not as the result of a legal provision, the resulting dose reports should also be stored long-term. Where they are worn as a legal requirement then the legal provisions should obviously be followed and these usually include specific record storage requirements. They usually require the records to follow the wearer, if he or she moves to another employer.

In the course of planning a survey, much data has to be collected on equipment and the site, including antenna sizes, mount heights, beam elevation settings, transmitter power, antenna gains and the like. This generally involves appreciable time and cost.

These data should be recorded on a card index or a similar computer record and linked to any changes made on the site so the record continues to reflect the actual status of the site. This is not only a wise safety provision but also tends to reduce the cost of preparation for surveys.

The triggering arrangement used to update the records is also a good method of automatically initiating an appraisal of the change by someone who is capable of determining whether a new survey is required and, if required, initiating it.

Training

General

Training in RF radiation safety, including the recognition of the coexistence of X-rays where relevant, may be needed for several different categories of people, for example:

- Design engineers, test engineers, operators and users who may or may not have technical backgrounds.
- Riggers and other mast and antenna structure workers.
- Maintenance and customer service engineers.

Although each group may have some unique requirements the training can be split into two basic categories:

1 People not required to have significant technical knowledge beyond their specific operating functions.
2 Technically trained people.

Category 1 is often neglected although all they often need is a simple understanding of the nature and effects of RF radiation, and information about safety provisions such as the purpose of safety shields and why they should not be removed without good reason; safety signs and prohibitions and other locally applicable arrangements for warnings and access limitation. Simple video films can be used for this sort of training, practical situations being illustrated directly.

RF safety training for technical people

For category 2 which encompasses technically trained people the basics of the training will be the same but with some particular emphasis for certain categories. For example, designers will be interested in design methods and techniques for RF safety and materials used for shielding and other purposes. Test engineers who are to undertake RF radiation safety testing and surveys will need more detail on measurement and on the instruments used together with practice in the use of those instruments and their limitations.

Experience in running training informally over many years and as formal courses over the last three years has brought to light a few useful points on courses where there is often a wide mix of senior and junior engineers with some safety practitioners from other disciplines such as mechanical engineering, medicine, transport, etc.

RF radiation measuring equipment is unfamiliar to many people and, as might be expected, the instruments are at some risk until they become more used to them in the course practical work.

Unless courses are run specifically for a particular organisation, courses attract a range of people from widely differing fields of RF work and training may involve explaining a little about different types of transmitters and RF radiation sources in use and providing some general information about the antenna systems used.

The basis of a course lasting two and a half days or three days [63] which is run monthly is outlined below. The aspects to be emphasised are:

1 Some of the course content needs to be adjusted to suit the experience and occupations of the students.
2 Military courses have an extra half day to cover various military organisational and medical aspects.
3 As much practical work as possible is put into the course, having regard to the problems of equipment availability, access to operating systems and the safety of other people at the site concerned. The practical work is geared to providing confidence in the correct use of instruments and understanding what to do in various circumstances rather than trying to provide a full experience in complete surveys.

The latter is impossible in the time and in any case, it is usually very difficult to get possession of equipment for the time required without disrupting work for other people. Some students such as doctors and senior engineers do not want training to the extent of doing full surveys as they will never do such work, but like to participate in something which gives them the feel of the working environment. In many cases, further experience in doing surveys, for those who will need to do such work, may have to be gained in the student's work environment.

The course typically covers:

1 General introduction to RF radiation; calculations for microwave antennas; calculations for the avoidance of risks to flammable vapours and electro-explosive devices.
2 Detailed consideration of the nature and effects of RF radiation; medical aspects; developments in the knowledge of RF radiation hazards.

3 RF safety standards and the background to them – ANSI/IEEE; INIRC/IRPA, EEC and national (NRPB in the UK case) standards and guides.

4 Theory and practical aspects of the use of RF radiation instruments; new instrument and monitor developments.

5 The nature of X-rays arising in RF radiation sources; legal requirements; measuring instruments and methods of measurement.

6 RF survey planning; typical data from surveys; methods of measurement; calculations for pulsed and non-pulsed transmissions, multiple irradiations, etc.

The practical work includes laboratory and field work. There is the chance to try various types of demonstration instruments loaned by the suppliers so that students are not limited to what is held locally.

The calibration of measuring equipment

The word 'calibration' is used both for the initial calibration of equipment and for subsequent calibration confidence checks. It is extremely important to have measuring equipment calibration checked at regular intervals in order to support the survey reports produced using that equipment. In some cases it will also be a legal requirement, particularly for ionising radiation measuring instruments. It may also be a condition of a quality standard to which the organisation operates e.g. MIL STD 45662[61] or the ISO 9000 series[64].

There are usually national sources of calibration for RF and X-ray instruments and instrument manufacturers should also be able to offer calibration traceable to national standards. There are some mutual agreements between national bodies for the acceptance of each other's national calibration certificates e.g. such as those between the USA, the UK and some other EEC countries, which facilitate trade and can reduce the effective costs involved in calibration when exporting.

Calibration of the instruments dealt with in this work are, typically, subjected to annual calibration. It has been found useful to plot the calibration results for each instrument.

By plotting a separate chart for each instrument but plotting cumulatively each time, i.e., successive calibration results for the same instrument on one chart, it is possible that trends can be identified.

For example, an instrument which seems to be drifting out of specification will show up. On the positive side, consistency provides confidence.

X-ray measuring equipment

Because of the lower energies likely to be met in dealing with transmitter X-ray measurements compared with the energies met in other fields such as X-ray machines, there is a need for low energy X-ray calibration checks and this can be done down to about 6 to 8 keV using fluorescent X-rays. Many countries have national or other calibration laboratories which can do this. In the UK the NRPB provides such facilities and will calibrate to the customers' requirements. The statutory calibration interval in the UK is 14 months.

RF radiation meters

The calibration of these instruments involves a very large capital expenditure on anechoic chambers, power sources and associated equipment. Consequently, such calibrations are inclined to be expensive.

They are often charged on the basis of so much per frequency, so they usually need a compromise between a large number of test frequencies with a large cost and few tests which are cheaper but give less confidence in the instrument performance.

In those laboratories which do a lot of calibration work, a degree of automation or at least the provision of aids to speed up the work is common. Figure 11.1 shows a picture of part of the calibration facility[60] at the UK National Physical Laboratory (NPL). The anechoic chamber is 5 m × 3 m × 3 m. It is used for the frequency range 2.45 GHz to 40 GHz.

A standard field is generated by feeding a known power into an antenna the gain of which is known. The equipment trolley is moved on accurately aligned rails. The instrument under test is located in the far field of the source antenna. The field can be calculated from the distance between the instrument under test and the source. The equipment is operated under the control of a computer with the RF power monitored by a calibrated directional coupler and power meter. The uncertainty in the standard field is better than ±0.5 dB (approximately ±12%).

At frequencies from 300 MHz to 2.4 GHz the calibration is done in a long tapered cell (Figure 11.2) with a field uncertainty of better than ±1 dB. For frequencies below 300 MHz Crawford type TEM cells are used. The overall frequency range for power density and field strength calibrations is 3 kHz to 40 GHz.

A number of RF radiation meter manufacturers offer calibration services, some of which conform to MIL Standard 455662[61]. A paper by Aslan[62] outlines the facilities used by the Narda company. These are broadly on the lines discussed above.

Often instrument specification limits are expressed in decibels rather than percentages and some people get confused with the conversion. Figure 11.3

Figure 11.1   *View of the 5 m calibration chamber used for the frequency range 2.45 to 40 GHz*
(Courtesy the National Physical Laboratory, Teddington, UK)

Figure 11.2 *TEM enclosures used for calibration in the range 3 kHz to 2.45 GHz*
(Courtesy the National Physical Laboratory, Teddington, UK)

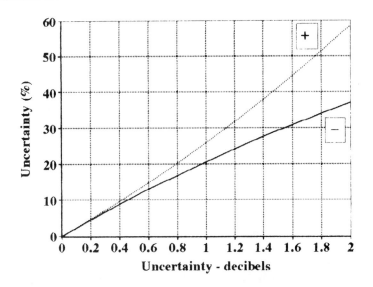

Figure 11.3 *Instrument uncertainties of measurement expressed in decibels and percentage*

provides a graph of percentage uncertainty against decibel uncertainty with one curve for positive values and one for negative values. For example an uncertainty of ±15% will give a decibel equivalence of +0.6 dB and −0.7 dB.

As new types of instrumentation are introduced, particularly limb current measurements and contact current measuring devices, there will be a demand for the calibration of these devices. Equally the progress along the RF frequency spectrum already results in power density measurement requirements with the consequential calibration requirements in excess of 100 MHz. These facilities are not widely available at present.

Some RF radiation safety instrument manufacturers also supply equipment such as TEM cells and other equipment for use in calibration facilities.

RF radiation incident investigation

There are a few important aspects of RF exposure investigations which can be identified as generally applicable. The greatest problems stem from the loss of information. The details that are most important are:

- Exactly where the subject was situated at the time.
- The source frequency, power, duty factor, modulation type, etc., if the source is known.

- The duration of the exposure; best checked with others present if possible, as people suddenly alarmed about being exposed may not be very accurate in estimating such things.
- The identity of other persons present.
- Details of what the subject claims – symptoms and any sensations felt; the location on the body, if specific such as a hot area or a burning feeling. These should be clearly recorded as claims and the agreement of the subject obtained to the written statement. If necessary, statements should also be obtained from people who were near to the incident at the time.

If a damaged component such as a cracked waveguide is involved, ensure that if it has been removed and replaced, the defective item is secured and retained.

The subject obviously needs to be medically examined promptly, in accordance with local procedures. However, unless there is something specific to look at e.g. a burn, reddening of the skin, optical damage or an acute condition such as hyperthermia, it is difficult for a doctor to relate RF exposure to changes in bodily condition. Often far too much is expected of doctors when the exposures are low and there is little to indicate anything of value about any consequences of the exposure.

At the initial examination the doctor will not have the benefit of the full report on the incident since it will not have been completed. He will need to see it when it is available. At the initial examination it is important to ensure that the doctor understands that RF radiation is involved. Cases have been recorded of such people being rushed to hospital and given a range of pointless tests for ionising radiation exposure. This could cause more stress than the original exposure!

Gathering information

The investigation may need to cover:
1 The exposure of the subject in power density or field unit terms, as measured in a simulation.
2 The nature of the exposure – whole body, or part exposure with a statement of the part affected e.g. arm, head, etc.
3 How the exposure arose – accident, breakage of a part, breach of established procedures, inadequacy of a procedure, absence of warning signs, etc.
4 The general awareness of safety requirements at the place where the exposure took place and the attitudes of the personnel and management concerned (effective safety management is very dependent on attitudes).
5 Any other form of radiation present such as ionising radiation, laser light, infrared or ultraviolet. These can easily be overlooked.

From the point of view of the investigator, the first need is to repeat the alleged incident situation with any faulty component replaced in the equipment and the same operating conditions as before. Obviously the investigation needs to be done very cautiously. Where calculations can be done they should be completed before any operational tests are done. For leakages this is often very difficult although simple calculations are possible for loose or open waveguides. If the component is too seriously damaged to be put back into the equipment, estimates will have to be made.

Before switching on the equipment, it is important to get the subject, or someone of similar build, to stand in the position where the exposure was obtained. It is then possible to make a sketch of the relationship of the likely radiation and the subject. Photographs are very useful if they can be arranged.

Investigations have been done by the author in a case where a male claimed irradiation of his genitals from an open waveguide. When the man was placed alongside the equipment it was found that the small leakage beam from the open waveguide on the low power equipment had a radius of about 10 cm at his body and was well above the navel.

Also, the level was within the standard applicable. This sort of inaccuracy is quite common in such a situation and implies a degree of panic on the part of the subject. It usually involves young men who are, perhaps understandably, worried about any possible harm in this region.

With the equipment radiating, a cautious survey recording levels in the area of interest is needed. If the levels expected significantly exceed permitted limits it may be necessary to reduce the power if possible, and scale up the consequent measurements results.

If the source is not known, probably a rare case, it is necessary to investigate the area in some detail, going back to the above actions when the source is established.

Reporting on incidents

The result of the survey should be a report giving details of the levels at a distance corresponding to that of the subject during the incident and at all parts of the body, not just the part mentioned by the subject.

Whilst the subject may have correctly identified the source equipment, it is important to allow for the possibility that he may be wrong both about the source and about the specific location of the leak. The surveyor should be careful not to be exposed due to such an incorrect judgement about the location of the radiation.

Where there are other sources of radiation nearby, it will usually be desirable to carry out measurements of them to avoid the possibility of other hazards co-existing and going undetected.

The use of diagrams in the report is important as it may be difficult for the report reader to visualise the equipment and the situation of the exposed subject. It has been said that one good diagram is better than several pages of text!

When the report is complete it can be assessed by the appropriate person. This is usually done by first comparing the results with the relevant standard. If the standard has not been exceeded then there is no need for any further calculations. If a standard has been exceeded, but only a part of the body has been exposed, it may be necessary to estimate the SAR for the affected parts of the subject's body. This is then compared with the standard to see whether such an exposure is permitted by any other clauses in the standard.

Reports should be clear, factual and frank. Where there are failures of the safety management system, it is usually better to define the failure in safety system terms so that any apportionment of individual blame is left to those who manage the people concerned. The fact that an exposure proves to be within permitted limits should not be considered as ending the matter if a safety system failure caused the exposure.

The final determination of any effects on the subject of the investigation will usually be the responsibility of the medical examiner or medical authority. There may also be other actions needed in the management of the activity if there are any adverse findings about safety provisions, safety procedures and disciplines and this may be the responsibility of the local management.

With frequencies under 100 MHz, there may be the additional complication of excessive induced currents in the body and it may be necessary to measure these. Since equipment to do such measurements is only just becoming available, experience is a little limited.

Finally, it should be noted that whilst undesirable exposures do occur from time to time, those having the responsibility for RF radiation safety are more likely to spend their time dealing with the psychological problems resulting from the secret worries of individuals about RF radiation which are fuelled from time to time by folklore and by the media, than with real incidents.

X-ray exposures

The foregoing relates to RF radiation incidents. Where an abnormal X-ray exposure is suspected or a combined RF and X-ray exposure is involved, the national provisions for ionising radiation incidents should be followed. This may involve formal reporting to national government safety organisations, as in the UK where it is covered by the Ionising Radiations Regulations[41].

References

1 Wood, J., Desert sounds; *IEE Review* pp. 275–80 July/August 1991.
2 Barker, A., Electromagnetic therapies – real or imaginary? *Physics World* January 1992 pp. G1–G2.
3 Saunders, R D et al., Biological effects of exposure to non-ionising electromagnetic fields and radiation: 3 *Radiofrequency and Microwave Radiation*; Report NRPB-R240 December 1991; ISBN 0-85951-332-7 HMSO Books, London.
4 IEEE Standard for safety levels with respect to human exposure to radio frequency electromagnetic fields; *IEEE* C95.1-1991; IEEE New York, USA. (Quoted in text by permission of the IEEE.)
5 Stuchly, S. S. et al., Energy Deposition in a Model of Man: Frequency Effects; *IEEE Trans. on Biomedical Engineering*, Vol. BME-33 No. 7 July 1986, pp. B1–B10.
6 Speigel, R. J. et al., Comparison of Finite-Difference Time-Domain SAR Calculations with Measurements in a Heterogeneous Model of Man; *IEEE Trans. on Biomedical Engineering*, Vol. 36 No. 8, August 1989 pp. 849–855.
7 Ellis, F. P. et al., *Physiological Responses to Hot Environments;* special report, Series No. 298. Medical Research Council, London pp. 158–79.
8 Mumford, W. W., Heat stress due to RF radiation; *IEEE* Vol. 57, No. 2, February 1969.
9 Adair, E. R., Thermophysical Effects of Electromagnetic Radiation; *IEEE Engineering in Medicine and Biology Magazine* March 1987 pp. 37–41.
10 Johnson, C. C., and Guy, A. W., *Non-ionising Electromagnetic Wave Effects in Biological Materials and Systems;* Proc. IEEE June 1972 Vol. 60 pp. 692–718.
11 Gandhi, O. P., and Riazi, A., Absorption of Millimetre Waves by Human Beings and its Biological Implications; *IEEE Trans. on Microwave Theory & Technology*, Vol. MTT-34 No. 2, February 1986 pp. 228–235.

12 Suess, M. J. and Benwell-Morrison, D. A., ed. *Non-ionising radiation protection*, 2nd Edition Copenhagen, WHO Regional Office for Europe, 1989 (WHO Regional Publications, European series No. 25). ISBN 92-890-1116-5. (Quoted in text with the permission of the World Health Organisation.)

13 Gandhi, O. P., *Advances in Dosimetry of Radiofrequency Radiation and their Past and Projected Impact on Safety Standards*; IEEE Proc. IMTC San Diego USA, April 1988 pp. 109–113.

14 Gandhi, O. P., *State of the Knowledge for Electromagnetic Absorbed Dose in Man and Animals*; Proc. IEEE Vol. 68, No.1, January 1980, pp. 24–32.

15 Frey, A. H., Auditory system response to radiofrequency energy; *Aerospace Medicine*, 32 December, 1961 p. 1140.

16 Frey, A. H., and Messenger, R., Human perception with pulsed ultra high frequency electromagnetic energy; *Science* Vol. 181, 27 July 1973 pp. 356–8.

17 Foster, K. R., and Finch, E. D., Microwave hearing: evidence for thermoaccoustic auditory stimulation by pulsed microwaves; *Science* 165, 1974 p. 256.

18 Gandhi, O. P., et al., Currents induced in a Human Being for plane-wave exposure conditions (0–50 MHz) and for RF sealers; *IEEE Trans. on Biomedical Engineering*, Vol. BME-33 No. 8, August 1986.

19 American National Standard ANSI C95.1-1982. *Safety levels with respect to RF electromagnetic fields 300 kHz to 100 HGz*. Quoted with permission of the IEEE.

20 Chen, J., and Gandhi, O. P., RF Currents induced in an anatomically-based model of a human for plane-wave exposures (20 to 100 MHz); *Health Physics* Vol. 57, No.1 (July) pp. 89–98 1989.

21 Williams, T., *EMC for product designers*, Butterworth-Heinemann 1992 ISBN 0-7506-1264-9.

22 Chatterjee, I., et al., Human Body Impedance and Threshold Currents for Perception and Pain for Contact Hazard Analysis in the VLF to MF Band; *IEEE Trans. on Biomedical Engineering*, Vol. BME-33, No. 5 May 1986, pp. 486–94.

23 *Electromagnetic Fields and the Risk of Cancer: Report of an Advisory Group on Non-Ionising Radiation*; NRPB Document Vol. 3, No. 1 HMSO Books, London ISBN 0-85951-346-7.

24 IEC 215 (1987) Safety requirements for radio transmitting equipment; IEC Geneva.

25 BS 3192:1987 *Safety requirements for radio transmitting equipment.* (Identical with IEC215 and EN60 215:1989) British Standards Institution.

26 Home Office, 1960. *Safety precautions relating to intense radio-frequency radiation*; HMSO Books, London.

27 INIRC 1988. Guidelines on limits of exposure to radiofrequency electromagnetic fields in the frequency range 100 kHz to 300 GHz; *Health Physics* 54, 115. 1988.

28 *Guidance as to restrictions on exposures to time-varying EM fields and the 1988 recommendations of the International Non-ionising Radiation Committee.* Document NRPB-GS11; HMSO Books, London.

29 Kowalczuk, C. I., et al., *Biological effects of exposure to non-ionising electromagnetic fields and radiation: 1. Static electric and magnetic fields; Chilton 1991.* Report no. NRPB-R238 HMSO Books, London.

30 Shinn, D. H., The avoidance of radiation hazards from microwave antennas; *The Marconi Review* Vol. 34, No. 201 2nd Quarter 1976. Note: References to BS 4992 therein are now invalid as BS 6656 superseded it.

31 BS 6656:1991 *Guide to the prevention of inadvertent ignition of flammable atmospheres by radio-frequency radiation*; British Standards Institution.

32 BS 6657:1991 *Guide to the prevention of inadvertent ignition of electro-explosive devices by radio-frequency radiation*; British Standards Institution.

33 Aslan, E., Broadband Isotropic Electromagnetic Radiation Monitor; *IEEE Trans.*, Vol. IM-21, No.4, November 1972 p. 421.

34 Hopfer, S., The Design of Broadband Resistive Radiation Probes; *IEEE Trans.* Vol. IM-21, No. 4, November 1972 p. 416.

35 Hopfer, S., and Adler, D., An ultra broadband (200 kHz–26 GHz) high-sensitivity probe; *IEEE trans.* Vol. IM-29, No. 4, December 1980.

36 *IEEE Standard recommended practice for the measurement of potentially hazardous electromagnetic fields*, IEEE C95.3-1991, IEEE New York, USA.

37 Mantiply, E. D., Characteristics of broadband radiofrequency field strength meters; *Proc IEEE Eng. in Medicine and Biology Society*, November 1988.

38 Radiation Measuring Hazard Systems: paper available from General Microwave Inc., describing the design of Raham instruments.

39 Blackwell, R. P. G., The personal current monitor: a novel ankle-worn device for the measurement of RF body current in a mobile subject; *J. Radiol. Prot.* 1990, Vol. 10, No. 2, pp. 109–114.

40 Systeme Internationale d'Unites; International System of Units; General Conference of Weights and Measures (CGPM).

41 *The Ionising Radiations Regulations*, 1985 (SI 1985 no. 1333); HMSO Books, London.

42 *Consultative Document NRPB-M321*: Board advice following publication of the 1990 Recommendations of ICRP, October 1991; HMSO Books, London.

43 Thomson Tubes Electroniques: *X–radiation from high power pulsed*

Klystron amplifiers reference NTH6111A June 1981. Quoted by permission of Thomson Tubes Electroniques, Paris.

44 Hunter B. E. X-ray Emission from Broadcast Transmitters: *IEEE Trans. on Broadcasting*, Vol. 36, No. 1, March 1990.

45 Philips Electronic Tubes catalogue; Philips Electronic Tubes.

46 Aslan, E., Non-ionising radiation – Measurement methods and artefacts; Loral-Narda Microwave Company, New York.

47 British Standard BS 5175:1976 *Specification for the safety of commercial electrical appliances using microwave energy for heating foodstuffs*; British Standards Institution.

48 USA Government Centre for Devices and Radiological Health (CDRH) – criteria and qualification requirements.

49 *United Kingdom frequency allocations chart*; Radiocommunications Division, Department of Trade and Industry, ISBN 0-11-514637-7; HMSO Books, London.

50 Guidance note PM51 from the UK Health and Safety Executive, *Plant and Machinery* 51; January 1986.

51 Narda informal paper 'Power density measurement on rotating beam radars (Burst peak measurements); Loral Microwave-Narda company 1983.

52 Aslan E. et al. Electromagnetic radiation measurements; *Microwave Product Digest* May/June 1989 pp. 30–2.

53 BS 4094 part 2: *Shielding from X-ray radiation*; British Standards Institution.

54 Manufacturer's specification for X-ray shielding glass reference RWB 46; Pilkington Special Glass, St Asaph, Clwyd, Wales, UK.

55 RFI shielding Ltd, Braintree, UK.

56 Ivco Vacuum Coatings, West Bromwich UK.

57 IEC562 1976 *Measurements of incidental ionising radiation from electronic tubes.*

58 BS 3510:1968 *Specification for a basic symbol to denote the actual or potential presence of ionising radiation*. British Standards Institution. (Technically equivalent to ISO361).

59 BS 5378 Part 3 *Safety signs and colours. Includes the non-ionising radiation sign*. British Standards Institution.

60 National Physical Laboratory (NPL) leaflet on power density and field strength calibrations; reference DES h 088 July 1990 NPL Teddington, UK.

61 MIL Standard 45662 1980 Calibration system requirements; Department of Defense, USA.

62 Aslan, E., An electromagnetic radiation monitor calibration in accordance with MIL STD 45662; *Journal of Microwave Power and EM energy*, Vol. 24, No. 2, 1989 pp. 102–7.

63 Course RFPO1, Marconi College, Chelmsford, UK.

64 ISO 9000 series *Quality System Standards* (BS 5750:1987 is identical).
65 Jokela, K, Theoretical and measured power density in front of VHF/UHF broadcasting antennas; *Health Physics*, Vol. 54, No. 5 (May) 1988 pp. 533–43.
66 *The Active Implantable Devices Directive*, EEC Commission, Brussels.
67 Allen, S. G. and Harlen, F., *Sources of exposure to radiofrequency and microwave radiations in the UK*; Report NRPB-R144 March 1983; ISBN 0-85951-195-2 HMSO Books, London.
68 The Electromagnetic Compatibility Directive (89/336/EEC); EEC Commission, Brussels.
69 Defence Standard 05-74/1 *Guide to the practical safety aspects of the use of radiofrequency energy*; MOD Glasgow, Scotland.
70 Hansen, R. C., Circular aperture axial power density; *Microwave Journal*, Vol. 19, February 1976 pp. 50–2.
71 BS 5501 part 1 1977 *General requirements: Construction and testing of electrical apparatus to ensure that it will not cause an explosion of the surrounding atmosphere.* British Standards Institution.
72 Scanlan, M. J. B., The design and measurement of large microwave antennas; *GEC Review*, Vol. 4, No. 2 1988 pp. 68–82.
73 Alexander, M. J., The design and performance of a large vertical aperture antenna for secondary surveillance radar; *GEC Review*, Vol. 4, No. 2 1988 pp. 109–117.

Sources of standards

ANSI C95 and IEEE Standards

IEEE Service Center
445, Hoes Lane,
P.O. Box 19539
Piscataway,
N. J. 08855 1331
USA

American National Standards Institute
1430 Broadway
New York NY10018
USA

British Standards

British Standards Institution
Sales Department
Linford Wood
Milton Keynes
MK14 6LE
United Kingdom

IEC Standards

(Also available through National Standards organisations)
IEC Central Office
1, Rue de Varambé
P.O. Box 131
1211 Geneva 20
Switzerland

Appendix 1 Technical and organisation abbreviations

Technical abbreviations

AC	Alternating current.
Am^{-1}	Magnetic field strength – amperes per metre.
DC	Direct current.
EED	Electro-explosive device.
EMC	Electromagnetic compatibility.
ERP	Equivalent radiated power.
r.m.s.	Root mean square value.
RF	Radio frequency.
RADHAZ	General term used for radiation hazards, particularly in the military field.
PFD	Power flux density; where used frequently this is usually shortened to 'power density' (PD).
RFI	Radio frequency interference (see EMC).
SAR	Specific absorption rate.
S	(In formulas unless otherwise noted) Power flux density.
Vm^{-1}	Electric field strength – volts per metre.
Wm^{-2}	Power flux density – watts per square metre.

Informal abbreviations used in this work

f.s.d.	Full scale deflection (of a meter).
DF	Duty factor (of a pulsed signal).
p.r.f.	Pulse repetition rate (Hz).
SF	Peak power density safety factor (overload protection for RF radiation instruments).

Modulation terms (may appear as upper or lower case)

AM	Amplitude modulation.

FM	Frequency modulation.
PM	Phase modulation.
SSB	Single sideband transmission.
ISB	Independent sideband transmission.
FSK	Frequency shift keying.
CW	Unmodulated carrier.

Organisations and systems

AGCIH	American Conference of Government Industrial Hygienists.
ANSI	American Standards Institution.
ATC	Air Traffic Control.
BNCE	British National Committee for Electroheat.
BSI	British Standards Institution .
BS	British Standard.
CAA	Civil Aviation Authority (UK).
CCIR	International Radio Consultative Committee.
CEN	European Organisation for Standardisation.
CENELEC	European Organisation for Electrotechnical Standardisation.
CISPR	International Special Committee for Radio Interference.
CCITT	International Telegraph and Telephone Consultative Committee.
DTI	Department of Trade and Industry (UK).
DEF STAN	UK Defence standards.
DRPS	UK Defence Radiological Protection Service.
EEC	European Economic Community.
EC	European Community.
EFTA	European Free Trade Area.
EN	European Norm (European standards).
ICRP	International Commission on Radiological Protection.
IEC	International Electrotechnical Commission.
IEEE	Institution of Electrical and Electronic Engineers (USA).
IEE	Institution of Electrical and Electronic Engineers (UK).
IRPA	International Radiological Protection Association.
INIRC	International Non-Ionising Radiation Committee.
ITU	International Telecommunications Union.
NAMAS	National Measurement Accreditation Service (UK).
NRPB	National Radiological Protection Board (UK).
NIST	National Institute for Science and Technology (formerly NBS) USA.
OJEC	Official Journal of the European Community.
SI	(1) Système Internationale.
	(2) Statutory Instrument (UK).

Appendix 2 Useful data and relationships

1 Quantities

Unit	Unit name	Symbol
Power, radiant flux	watt	W
Power density, irradiance	watt per sq. metre	Wm^{-2}
Electric field strength	volt per sq. metre	Vm^{-1}
Magnetic field strength	ampere per sq. metre	Am^{-1}
Magnetic flux density	tesla	T
Specific energy	joule per kilogram	Jkg^{-2}
Electric resistance	ohm	Ω
Electric conductance	siemen	S
Electrical potential, Electromotive force	volt	V
Electric current	ampere	A
Frequency (cycles or repetitive events per second)	hertz	Hz

2 Units

Non-SI and SI unit relationships

1 kilowatt-hour = 3.6 MJ

1 hp (horsepower) = 745.7 W

1 inch = 25.4 mm

1 foot = 0.305 m

1 yard = 0.914 m

1 mile = 1.609 m

1 mm = 0.039 inch

1 cm = 0.394 inch

1 m = 1.094 yd

1 km = 0.621 mile

1 metre per second = $3.281\,\mathrm{fts}^{-1}$

$1\,\mathrm{fts}^{-1} = 0.305\,\mathrm{ms}^{-1}$

Constants and other relationships

Velocity of light in free space $= 2.997\ 925 \times 10^8\ \mathrm{ms}^{-1}$ (generally treated as being $3 \times 10^8\ \mathrm{ms}^{-1}$)

Electron volt (eV) $= 1.602 \times 10^{-19}\ \mathrm{J}$
1 microtesla (μT) $= 10$ milligauss (mG)
$1\ \mu$T $\quad\quad\quad = 0.8\ \mathrm{Am}^{-1}$
$1\ $mG $\quad\quad\quad = 0.08\ \mathrm{Am}^{-1}$

3 Old and new ionising radiation units

Old units

Absorbed dose	rad (rad)	$1\ \mathrm{rad} = 0.01\ \mathrm{Gy}$
		$1\ \mathrm{Gy} = 100\ \mathrm{rad}$
Dose equivalent	rem (rem)	$1\ \mathrm{rem} = 0.01\ \mathrm{Sv}$
		$1\ \mathrm{Sv} = 100\ \mathrm{rem}$
Exposure	röentgen (R)	$1\ \mathrm{R} = 2.58 \times 10^{-4}\ \mathrm{Ckg}^{-1}$

(This unit is no longer used and there is no SI replacement.)

When converting to the new units, rads, rems and röentgen are considered to be numerically equal.

New SI units

| Absorbed dose | gray (Gy) | $1\ \mathrm{Gy} = 1\mathrm{Jkg}^{-1}$ |
| Dose equivalent | sievert (Sv) | $1\ \mathrm{Sv} = 1\mathrm{Jkg}^{-1}$ |

Dose equivalent = absorbed dose multiplied by the quality factor (QF) and QF for X-rays and gamma rays $= 1$. Hence in this case the absorbed dose and the dose equivalent are numerically equal.

Power density and the electric/magnetic field relationship

Figure A2.1 provides a quick reference graph for the power density and electric field relationship for plane wave conditions, covering the range 1 to $1000\ \mathrm{Wm}^{-2}$. If converting mWcm^{-2}, multiply by ten first to convert to Wm^{-2}.

Figure A2.2 is a similar quick reference graph for power density and magnetic field covering the range 1 to $100\ \mathrm{Wm}^{-2}$.

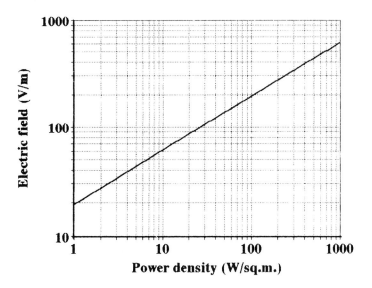

Figure A2.1   *Graph of values of power density (Wm⁻²) versus electric field plane wave values (Vm⁻¹)*

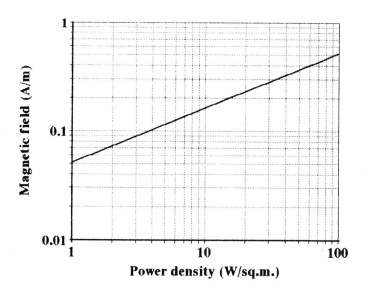

Figure A2.2   *Graph of values of power density (Wm⁻²) versus magnetic field plane wave values (Am⁻¹)*

Index